George Kennedy
'66

Electronic Servicing Data and Procedures:

A Complete Manual and Guide

ROBERT C. GENN, JR.

Illustrated by E. L. Genn

Prentice-Hall, Inc. Englewood Cliffs, New Jersey
(Business and Professional Division)

Prentice-Hall International, Inc., *London*
Prentice-Hall of Australia, Pty. Ltd., *Sydney*
Prentice-Hall Canada, Inc., *Toronto*
Prentice-Hall of India Private Ltd., *New Delhi*
Prentice-Hall of Japan, Inc., *Tokyo*
Prentice-Hall of Southeast Asia Pte. Ltd., *Singapore*
Whitehall Books, Ltd., Wellington, *New Zealand*
Editora Prentice-Hall do Brasil Ltda., *Rio de Janeiro*
Prentice-Hall Hispanoamericana, S.A., *Mexico*

© 1986 *by*
Prentice-Hall, Inc.
Englewood Cliffs, New Jersey

First Printing

Editor: George E. Parker

Library of Congress Cataloging-in-Publication Data
Genn, Robert C.
  Electronic servicing data
and procedures.

  Includes index.
  1. Electronic apparatus and appliances—
Maintenance and repair. I. Title.
TK7870.2.G44 1986   621.381′028′8   85-12344

ISBN 0-13-251851-1

Printed in the United States of America

How This Book Will Help You Perform Virtually Any Electronic Servicing Job

Here—completely revised, up-dated and enlarged—is a *modern handbook* of electronic servicing data, tests and measurements that will enable you to solve quickly virtually any electronic repair problem found in the average service shop. You'll learn how to make and analyze the results of *almost two hundred tests and measurements,* as well as how to come to the correct conclusions that are absolutely essential for successful troubleshooting.

Each section takes you step-by-step through simplified servicing data and procedures that are of real practical value to any technician. In Section 1, audio equipment tests and measurements are analyzed in detail. There are over 20 step-by-step test procedures included.

The second section contains an analysis of all common digital equipment tests and how to make them. Over 25 example test guides and procedures for all types of digital ICs will be found in this section.

Next, comprehensive guidelines for solid state equipment servicing, including microprocessing units (MPUs) and computer systems, are given in Section 3. Again, you'll find many tests and measurement guides explained.

Section 4 includes a wide range of practical servicing information, techniques and shortcuts based on actual shop experience. An explanation of how to test modern rf systems, and particularly, radio receivers is included. This section contains analyses of all common radio receiver tests and how to make them—everything from the antenna to the audio stages.

Getting maximum performance from electronic servicing gear is a never-ending quest for all successful technicians. To do this, means selecting the test instrument that is best for the particular measurement and then performing each test in the most efficient manner.

You'll find the factors to be considered for hundreds of servicing jobs in this handbook. For example, servicing solid state TV receivers can be a frustrating experience because special parts such as Thyristors, ICs, and "super modules" are encountered more and more. To make your servicing easier, there are scores of both digital and analog tests and measurements to help you service both today's and tomorrow's modern solid state TV receivers.

Here are just a few of the many tests and measurements and shop hints this book contains:

1. An example of how to check optoisolators, phototransistors, a light-operated relay, logic gate ICs, and dozens of other ICs.

2. Practical CB radio (AM and SSB) servicing tests and measurements.

3. How to build a logic memory probe, logic pulser, logic monitor, and how to test them.

Every section shows you how to get the most out of your tests and instruments. All state-of-the-art ICs are explained in detail showing example pin numbers and input/output signals. You will also learn how to make reliable measurements with your VOM, VTVM, and scope that you may not have believed possible. For example, how to use your VOM's low-current range to measure low voltages. You can be assured *every* important phase of practical electronic servicing is covered.

In the past, we needed to remember only a few analog servicing techniques to service almost every piece of electronic gear manufactured. But digital integrated circuits, on-chip microcomputers, optoelectronics, and many other solid state devices have changed all this. Today, we must service an incredible variety of display and solid state circuits, plus all of their associated components, creating the need for a modern encyclopedic handbook of electronic servicing such as this one.

As you can see, this handbook is a *practical* treasure of "real life" servicing techniques. It will quickly become one of the most valuable everyday aids you've ever used.

Robert C. Genn, Jr.

Other Books by the Author

The Complete Microcomputer Handbook—With Tested
    Basic Programs
Illustrated Guide to Practical Solid State Circuits—With
    Experiments and Projects
Digital Electronics: A Workbench Guide to Circuits, Experi-
    ments and Applications
Practical Handbook of Solid State Troubleshooting
Manual of Electronic Servicing Tests and Measurements
Practical Handbook of Low-Cost Electronic Test Equipment
Workbench Guide to Electronic Troubleshooting

CONTENTS

Section 2

PRACTICAL GUIDE TO MAINTAINING DIGITAL EQUIPMENT

Section 3

PRACTICAL GUIDE TO SERVICING SOLID STATE CIRCUITS AND DEVICES

Section 4

PRACTICAL RADIO FREQUENCY EQUIPMENT TEST AND MEASUREMENT GUIDE

Section 5

PRACTICAL GUIDE TO IN-CIRCUIT TESTING OF HIGH FREQUENCY TRANSISTORS

Section 6

TRANSMITTER/RECEIVER SERVICING

Section 7

TV SERVICING TESTS AND MEASUREMENTS

Section 8

TESTS AND MEASUREMENTS FOR ANTENNA SYSTEMS AND TRANSMISSION LINES

Section 9

PRACTICAL TESTING, MEASURING AND DIGITAL INSTRUMENT-BUILDING TECHNIQUES

Section 10

TIMESAVING TESTS AND MEASUREMENTS FOR SEMICONDUCTORS

SECTION 1

Audio Equipment Tests and Measurements

SECTION 1.1: A SERVICEMAN'S GUIDE TO THE BASIC PRINCIPLES OF SOUND

Cutting servicing time when working with audio equipment begins with developing and understanding of the basic principles and behavior of sound generation. Sound waves are longitudinal *mechanical* waves. For us at the workbench, mechanical is probably the most important word in the sentence because it means sound can only be propagated in solids, liquids, and gases.

Actually, there is a large range of frequencies within which this type of wave can be generated. However, as a general rule, when working on audio equipment we are only interested in the *audible* range (about 20 Hz to 20,000 Hz). Incidentally, the audio frequency waves below the audible range are called *infrasonic waves* and above, *ultrasonic waves.*

For our purposes, it's better to define sound waves as a wave motion that is propagated in air after originating in vibrating strings (human vocal cords, violins, etc.), and vibrating plates (loudspeakers, drums, etc.).

In general, at the workbench we are primarily concerned with electrical and electronic processing of these sound waves. Nevertheless, it is important for you to realize that the original audio waveforms should be approximately periodic or consist of a small number of approximately periodic components (for example, musical sounds). If not, the electronic processing equipment (high-fidelity system) will produce sounds whose waveforms are very irregular and will be heard (or measured) considerably distorted. Or, to put it another way, you should have a good quality stereo signal generator; i.e., the generator should have a very low percentage of distortion.

What do we mean when we say audio distortion? To put it simply, distortion is any unwanted change in an audio waveform such as harmonics combining with the desired frequency waveform. See Section 1.4 for a complete analysis of distortion and its measurements, etc.

1

One specification you will find given for almost every audio system is "Total Harmonic Distortion" (usually called *thd*). In order to measure thd you must apply a signal with much better quality than the amplifier being tested. This is why we emphasized that a good quality stereo generator is a basic requirement for an audio shop. You will find more about thd analyzers later in this section.

At this point we should point out that noise, interference, and hum are not considered forms of distortion. Noise is a random sound composed of many different frequencies *not harmonically related*. You will find that most technicians agree that high-fidelity reproduction should have a frequency response that is uniform within at least ± 1 dB for 20 Hz to 20 kHz (or better) with a thd less than 1 percent at any frequency within that range.

Many audio systems do not utilize high-fidelity reproduction over the entire audible range. For example, the reproduction of speech may be limited to a bandwidth of 250 to 3,500 Hz with a high degree of intelligibility. Tests have shown that if you remove the higher frequencies (using a filter), you'll find that at 1,550 Hz intelligibility falls off to 65 percent. From about 800 Hz up, lies the greater portion of human speech intelligibility. As a rule, the lower frequencies aren't that important (in reference to intelligibility). The human ear is actually less sensitive at both the lower and upper ends of the audible frequency spectrum and even this characteristic usually varies with both sex and age.

SECTION 1.2: AUDIO AMPLIFIER DESIGN BASICS
FOR THE REPAIRMAN

Look at almost any retailer's catalog—for instance, Radio Shack's—and you will quickly realize that audio technology is one of the rapidly changing areas in electronics. High-fidelity receivers, in particular, have evolved greatly within recent years. For example, many of today's receivers are all-digital, quartz-locked, and include station memory. Also, they have 50 watts (rms) minimum per channel into eight ohms, from 20 to 20 kHz, with no more than 0.02 percent thd. Analysing design principles (and specifications) for systems and components such as these, is covered in this section. An example of a linear amplifier integrated circuit (IC) is shown in Figure 1-1.

Another solid state, low power audio amplifier that you might encounter is an MC1306P. This is a 1/2-watt complementary preamplifier and power amplifier designed in a single 8-pin dual-in-line package. Figure 1-2 is a schematic of a phonograph amplifier, designed around this IC, using a ceramic cartridge on the input.

You will encounter many audio amplifier designs that use a set of complementary power transistors; for example, the 2N6111 (PNP) and

(A)

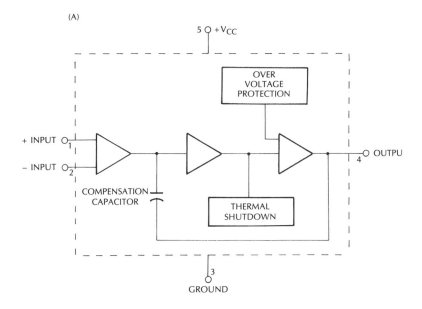

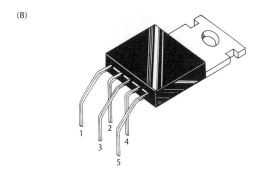

Figure 1-1: An example of an 8-watt audio power amplifier (Motorola's TDA2002) designed for automotive and general purpose applications.

2N6288 (NPN) might be used. Figure 1-3 shows the package used for both of these transistors.

A set of transistors such as these two usually uses one of the basic output-tranformerless (OTL) types of audio output circuit designs. Figure 1-4 shows a simplified diagram of a complementary symmetry amplifier using these two transistors.

Connecting two transistors in this manner (see Figure 1-4), causes each transistor to conduct over one-half of the input cycle because, as you can see, Q_1 is a PNP and Q_2 is a NPN. Recalling the days when you were studying basic

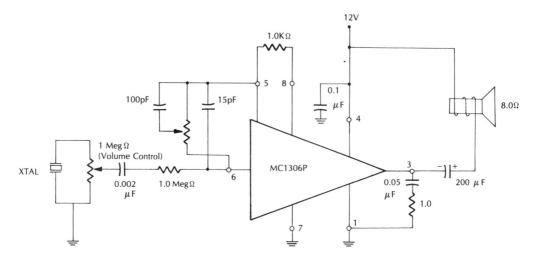

Figure 1-2: Phonograph amplifier design using Motorola's MC1306P IC.

Figure 1-3: The type of case you would encounter in an audio amplifier using 2N6111 and 2N6288 complementary power transistors.

electronic theory, you'll realize that this is simply a Class B amplifier design principle and, therefore, fairly easy to troubleshoot. *Note:* Transistors such as these two (40 watts), usually require small fans or blowers, plus heat sinks for cooling. Another, and very popular, OTL amplifier (a stacked circuit) that you might run into is shown in Figure 1-5.

An example of a Darlington pair, also called *Darlington amplifier, double-emitter amplifier,* and sometimes β *(beta) amplifier* is shown in Figure 1-6. Power transistors such as the ones shown in Figure 1-6 (a 40-watt power transistor), must be mounted on a heat sink and usually require forced air cooling before making any type of operational check. Figure 1-7 shows a pair of Darlingtons used to form an OTL arrangement. This circuit is very good because the percent of distortion is very low.

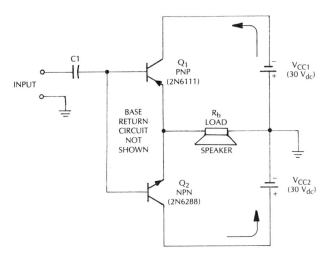

Figure 1-4: A stripped-down circuit diagram for a complementary amplifier you might encounter.

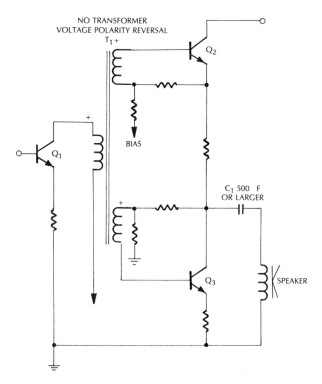

Figure 1-5: An OTL amplifier design, often called a *stacked circuit*, that is often used in audio systems.

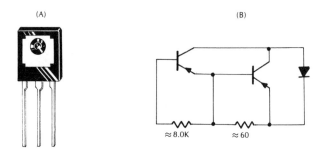

Figure 1-6: A PNP Darlington complementary amplifier (2N6034).

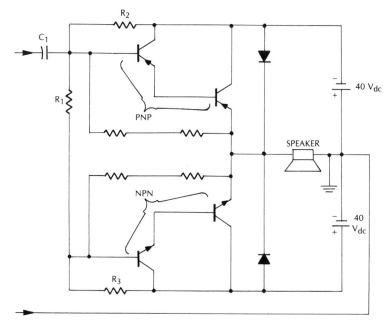

Figure 1-7: Darlington OTL circuit.

SECTION 1-3: HOW TO ANALYZE AUDIO AMPLIFIER DESIGNS

In general, most of the basic information you'll need to analyze a particular audio amplifier design can be obtained from the data sheet for the solid state device (transistor, IC, etc.) the amplifier is using.

Two questions that are important in the beginning of your analysis are 1) what is the amplifier output power? and, 2) what is the speaker impedance that the amplifier must drive? With this information you can quickly calculate the approximate power supply voltages and various circuit component values.

As an example, let's assume that you have an OTL amplifier that drives a 4-ohm speaker (two 8-ohm in parallel) and the amplifier is rated at 8 watts. To find the maximum current through the operating transistors that will develop 8 watts in a 4-ohm load, we can use the formula $I_C = \sqrt{P/R}$ (remember, only one transistor operates at a time in an OTL circuit). The maximum collector current (I_C) is

$$I_C = \sqrt{P/R} = \sqrt{8/4} = \sqrt{2} = 1.414 \text{ A (rms)}.$$

The peak current through the transistor would be

$$I_{max} = 1.414 \times I_C = 1.414 \times 1.414 = \text{approx. 2A}.$$

If you are using a peak reading voltmeter, you should find about 8 volts when the peak current (2A) is flowing through the 4-ohm speaker; that is, about 8 volts maximum is the voltage drop across each of the transistors. Incidentally, the power supply usually needs to produce about twice the transistor voltage and should read about 20 V.

Now, if you know the beta of the transistor, you can calculate the approximate base current (I_b). Assuming a beta of 40, we have

$$I_{bmax} = I_{Cmax}/\beta = 2/40 = 50 \text{ mA}.$$

However, to prevent clipping, it is common practice to increase this base current by about 50 percent. If this isn't done, you could end up with a value of 75 mA.

With all this information, it is fairly simple to calculate almost every component value in the circuit by using Ohm's law. The only other component that might give you trouble is the output coupling capacitor. The minimum value you would want to use between the amplifier output and speaker can be found by using the formula

$$C = 1/[(2\pi)(R_L)(f_{min})] \qquad (\text{for } -3\text{dB at } f_{min})$$

where: $2\pi = 6.28318$

   R_L = speaker impedance

   f_{min} = lowest frequency that will produce 1/2 power (-3 dB).

In our example, using the lowest standard audio (20 Hz), we arrive at a value of

$$1/(6.28318 \times 4 \times 20) = \text{approx. 2000 } \mu\text{F}$$

At the workbench you may find output coupling capacitors with a lower value such as 1000 μF. Smaller values do cause a slight loss at the low end of the audio range, however.

A few words about power dissipation. In a Class B amplifier such as we are discussing, each power transistor should be able to dissipate slightly more than 1/4 the total output power. The output power in our example is 8 watts, so we need a pair of transistors that are capable of dissipating at least 2 watts (preferably more).

SECTION 1.4: DISTORTION IN AUDIO SYSTEMS

As we promised in Section 1.1, a comprehensive coverage of distortion is presented in this section, beginning with Total Harmonic Distortion.

What does it mean when we read a catalog and it claims that the receiver in the advertisement has no more than a certain percent (for example, 0.02) of total harmonic distortion? Actually, one of the chief forms of distortion is frequency (the others are amplitude and phase). As an example, when we say total harmonic, we are automatically considering frequency. This is because we are discussing a certain fundamental frequency measured at the output of some system (audio amplifier, receiver, etc.) as a ratio to the power of all harmonics measured at the output of the system.

Harmonic distortion meters contain audio frequency filters for signal processing. Thus, the meter employs a sharp-tuned filter that eliminates the fundamental frequency from the signal under test. In turn, if there are any harmonics they are passed on to an ac voltmeter that is usually calibrated in harmonic distortion percentage units. A test such as this one is ordinarily made at 1 kHz, but you can make the test at any frequency. However, remember that an amplifier will produce different readings at different test frequencies.

SECTION 1.5: TESTING AUDIO SYSTEMS FOR HARMONICS

There is a simple way to check for harmonics. Use the circuit shown in Figure 1-8 and follow the procedure given.

Test Equipment:
AC voltmeter, capacitor, resistor, and the audio sine-wave generator or amplifier under test.

Test Setup:
See Figure 1-8.

Comments:
This test is based on the fact that the voltage reading (E_1) will be an exact 45° diagonal of the voltages E_2 and E_3, when plotted, if $X_C = R$ *and there are no harmonics.*

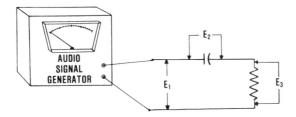

Figure 1-8: Equipment setup for testing an audio sine-wave generator for harmonics.

Procedure:

Step 1. Select the values of R and C that will produce approximately the same voltage readings at your chosen frequency.

Step 2. To compute the value of the capacitor, simply use the formula C = 0.159/(F$_r$X$_c$), where F$_r$ is the frequency you chose to use, and X$_c$ is the same value as the resistor.

Step 3. After you've made all three voltage readings, draw a plot like the one shown in Figure 1-9.

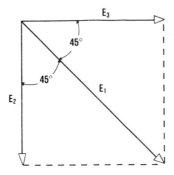

Figure 1-9: A plot similar to the one shown indicates the audio signal generator is free of harmonics.

If you find E$_2$ and E$_3$ are equal in length and the E$_1$ length completes the diagonal line, it's an indication that the audio signal generator or amplifier is free of harmonics. But if you can't plot a rectangle as shown, the signal source is producing harmonics.

SECTION 1.6: OUTPUT POWER VERSUS POWER SUPPLY VOLTAGES USING EITHER 8- OR 16-OHM SPEAKERS

It should be pointed out that several things can affect the thd of an audio amplifier; i.e., the output power, the supply voltage, and load (speaker) you place on the amplifier during the test.

For example, let's say that we decide to use a 16-ohm speaker during a test of an amplifier designed to drive an 8-ohm speaker. By referring to Figure 1-10, it's apparent that the first thing we will probably notice is a decrease in output power, regardless of where we set the supply voltage.

Notice, it is important that we use the correct driving power, load, and supply voltage during a test, or we will pay for it with a higher thd. To put it another way, if the thd meter reads abnormally high for a certain audio system, check the speaker impedance (see if it is correct for the system under test), check for high or low voltages, and check to see if you are overdriving the circuit.

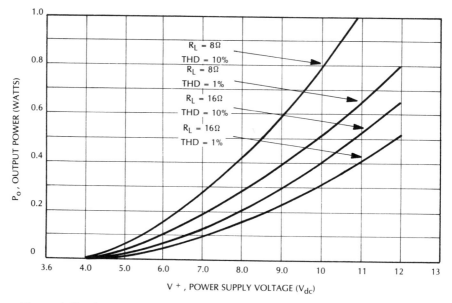

Figure 1-10: Output power versus power supply voltages using 8- or 16-ohm speakers. Total harmonic distortion is also shown for the two loads at various voltages and output powers.

SECTION 1.7: HOW TO MEASURE INPUT RESISTANCE (ATTENUATOR PAD)

Simple Ohmmeter Method

Test Equipment:

Ohmmeter and terminating resistor.

Test Setup:
    See **Procedure.**

Comments:
    There are several different types of attenuator pads that can be checked with this procedure. Figure 1-11 shows some basic attenuator pads.

Procedure:
    Step 1. Connect a terminating resistor of equal value to the pad's normal load, across the pad's output. For example, if the pad is connected between two 75-ohm lines during normal operation, use a 75-ohm resistor (the power rating of the resistor isn't important).

    Step 2. Measure the input resistance of the pad with your ohmmeter. The resistance you measure should equal the normal operating impedance of the pad.

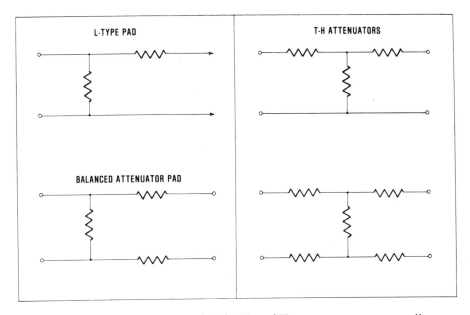

Figure 1-11: Basic attenuator pads. The H- and T-type attenuators generally are used only to attenuate audio signals and not for impedance matching. Therefore, their input and output impedances usually are equal.

Voltmeter and Audio Signal Generator Method

Test Equipment:
    Audio frequency generator, high input impedance voltmeter, load resistor with a resistance value equal to the output impedance of the device under test, and an adjustable precision resistor (a resistor decade box is the easiest to work with).

Test Setup:

See Figure 1-12.

Procedure:

Step 1. Connect the test setup as shown in Figure 1-12. The variable resistor R_1 is *not* connected in series with the audio frequency generator until you reach Step 4.

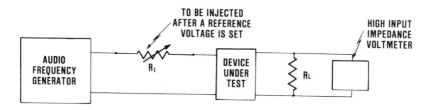

Figure 1-12: Test setup for measuring the input resistance of any resistive device having an input resistance under 10 k Ω.

Step 2. Set the audio signal generator to operate at 1,000 Hz.

Step 3. Adjust the output level of the signal generator until you have a convenient ac voltage reading on the voltmeter—about 5 or 10 volts—without R_1 in the circuit.

Step 4. Insert R_1 into the test circuit and adjust its value until the voltmeter reading is exactly one-half of what you had in Step 3.

Step 5. Measure, or read, the value of R_1. For all practical purposes, the value of R_1 is equal to the input resistance of the device under test.

SECTION 1.8: ATTENUATOR REACTANCE RESPONSE TEST

Test Equipment:

Audio signal generator, voltmeter, load resistor, and variable resistor.

Test Setup:

See Figure 1-12.

Procedure:

Use the same procedure explained in the last test for all frequencies of interest. At some upper frequency, the input resistance should begin to drop to a lower value due to reactance shunting. This is the point where the circuit's reactance is beginning to affect your reading; i.e., the circuit is no longer predominantly resistive.

SECTION 1.9: VARIABLE ATTENUATOR PAD TEST

Test Equipment:
　　Ohmmeter and two load resistors having the same resistance values as the variable attenuator input and output impedance.

Test Setup:
　　First, connect your ohmmeter to the input (or output) of the pad and connect the proper value load resistor to the opposite end. The power rating of the resistor is not important. Simply reverse the setup to make the second check.

Comments:
　　Many constant impedance attenuators are variable. In this case, simply adjust the step attenuator or variable control to all possible settings. The ohmmeter reading should remain essentially the same for all positions (within the tolerance of the pad). Figure 1-13 shows a simple schematic of a variable T-pad and a bridged T-pad.

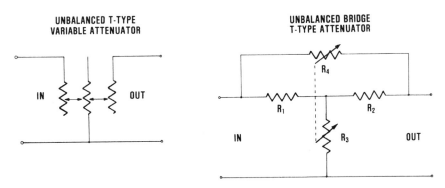

UNBALANCED T-TYPE VARIABLE ATTENUATOR

UNBALANCED BRIDGE T-TYPE ATTENUATOR

Figure 1-13: Continuously variable attentuators that can be checked with an ohmmeter.

　　It doesn't make any difference to which end of the pad you connect your ohmmeter, to begin with (in or out). However, it is important that you connect the proper value of resistance—it must be equal to the line impedance—to the opposite end of the pad or your readings may be in error.

Procedure:
　　Step 1. Connect the proper value of the load resistance to the output of the constant impedance variable attenuator.
　　Step 2. Connect your ohmmeter across the input terminals of the pad.
　　Step 3. Vary the attenuator control knob, or switch, through its range, at

the same time watching the ohmmeter reading. The reading should remain almost constant (some pads have a little more tolerance than others, therefore you might see a slight meter variation in some cases).

Step 4. Reverse the setup and connect a load resistor having the same resistance value as the pad's input impedance.

Step 5. Connect the ohmmeter to the pad output terminals. Again, watch the ohmmeter reading as you vary the attenuator control. The reading should remain essentially constant. If there is little or no change in the resistance reading, the pad can be assumed to be properly matched to the lines.

SECTION 1.10: FREQUENCY RESPONSE TESTS

Sine-Wave Generator and AC Voltmeter Method

Test Equipment:
        AC voltmeter, sine-wave generator and terminating resistor.

Test Setup:
        See Figure 1-14.

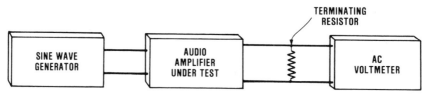

Figure 1:14: Test setup for frequency response measurement.

Comments:
        Some signal generators don't have a metered output. In this case, you can use a second ac voltmeter to maintain a constant output level. However, to prevent erroneous readings, both meters must have the same frequency response. Also, to prevent distortion, and for the safety of the system, place a dc blocking capacitor between the signal generator and system, and use a value greater than the value for the capacitor in the input circuit (for example, the coupling capacitor connected to the base lead of a common emitter bipolar transistor amplifier).

Procedure:
        Step 1. Start off with about 1/10 the highest normal operating power during warmup—about 30 mintues to 1 hour—and it's important you have a stable ac power line voltage (1 percent is considered excellent). Next,

assuming the equipment is operating with no overload, 1/3 maximum voltage output, low distortion, and has a fairly flat frequency response, make the frequency response measurement.

If you're not sure the equipment has a comparatively flat frequency response, simply sweep the signal generator across the band and watch the ac voltmeter for a maximum reading. If there isn't much difference in readings, use 1 kHz for reference when you make the plot. The sketch in Figure 1-15 shows a typical frequency versus voltage output plot.

Should there be a substantial difference in readings when you sweep the equipment bandpass, use the frequency with the highest voltage reading as your reference. This is to prevent overdriving the system.

Step 2. Apply a constant amplitude 1 kHz signal voltage that will produce an output signal about 1/3 the maximum voltage output to the input of the equipment under test.

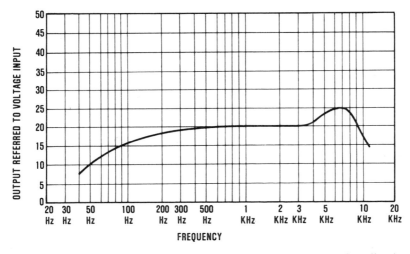

Figure 1-15: A typical frequency curve made using voltage and audio signal generator frequency settings with a reference of 1 kHz.

Step 3. Measure the system output voltage at several different frequencies (for instance, 60, 240, 1050, 3500, and 10,000 Hz) with the ac voltmeter.

Step 4. Make a plot of the output versus frequency and this is your frequency response curve.

Sine-Wave Generator and Oscilloscope Method

Test Equipment:

Sine-wave generator, oscilloscope and terminating resistor, if needed, (refer to manufacturer's service notes)

Test Setup:

Connect the sine-wave oscillator to the input terminals of the equipment under test, and the oscilloscope to the output terminals, using a high impedance probe on the scope.

Comments:

Another approach to testing audio equipment frequency response is to use a sine-wave oscillator and ordinary oscilloscope (a dual-or-more trace would be better for the following tests), equipped with a high impedance probe. It's possible to connect the scope directly; however, a high impedance probe is better in almost every case. The procedure is simple. To check the quality of the system, merely inject the sine-wave into the input circuit and view its output on the scope.

Procedure:

Connect the scope test leads to the equipment output and adjust its sweep frequency (time base) and other controls until you have two or three cycles appearing on the scope display. If the equipment has a good frequency response, you'll see an excellent reproduction of the input sine wave at any frequency from about 50 Hz to 20 kHz ... assuming you are testing hi-fi equipment. Poor performance will be indicated by clipping of positive peaks, negative peaks, or both. Some clipping is sure to occur, but if the sine wave appears basically the same for all test frequencies, the equipment is considered to have a flat frequency response. In Figure 1-16 you'll find some of the various scope presentations that indicate a system is not responding properly. However, the oscilloscope must be free of distortion to reproduce a perfect copy of the test signal.

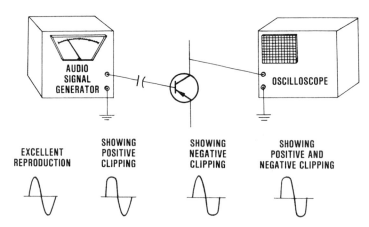

Figure 1-16: Distorted sine-wave patterns that may be seen on an oscilloscope, which indicate an audio system is not responding properly during a frequency response check.

Square-Wave Generator and Oscilloscope Method

Test Equipment:

Square-wave generator and oscilloscope utilizing a high impedance probe

Comments:

Deviations which may be seen on the scope that indicate specific problems in the frequency response of an amplifier are shown in Figure 1-17. At (A), the observed waveform indicates a loss in low frequency response. The problem may be an improper value (too small) coupling capacitor. Capacitors always decrease in value if they become partially open. At (B), the waveform shows a substantial drop in high frequency response. If you observe this pattern at certain high frequencies, the eighth harmonic of a 2 kHz test signal for instance, the amplifier can be considered to be linear up to 16 kHz (8 × 2 kHz = 16 kHz). The waveform at (C) indicates there is a higher gain at the lower frequencies. On the other hand, because the lower frequencies determine the shape of the horizontal portion of the square wave, it will sag in the middle if there isn't sufficient amplification at these frequencies, as shown in (A).

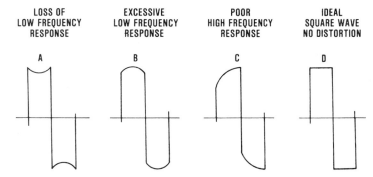

Figure 1-17: Scope presentations of frequency response deficiencies indicated by a change in a square-wave test signal.

SECTION 1.11: LINEARITY TEST PROCEDURE

To make a proper linearity test, we first must know what we are trying to verify. Basically, all we want to determine is at what range is the amplifier output signal always a replica of the input signal. This relationship generally exists only over a limited range of signal input voltages and often only in a select band of frequencies (for example, the audio band, 20 to 20 kHz).

It should be pointed out that there is a basic distinction between the mixing of one audio waveform with another and the modulation of one audio waveform by another. Audio waveform mixing is termed a *linear* process, whereas audio waveform modulation is termed a *nonlinear* process. The following tests determine the linearity of an amplifier, not such things as intermodulation, etc.

Sine-Wave Generator and Oscilloscope Method

Test Equipment:
   Sine-wave signal generator and oscilloscope.

Test Setup:
   See Figure 1-18.

Comments:
   To use a scope to make an audio amplifier linearity test, the first step is to check the scope linearity. Although this should be done in almost all cases, it is a must in this test because a reference is required in order to make a comparison when evaluating the amplifier. To check your scope, connect the output test lead of the signal generator to both the vertical and external horizontal inputs of the scope, as shown in Figure 1-18.

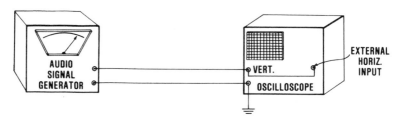

Figure 1-18: Test setup for checking a scope amplifier's linearity.

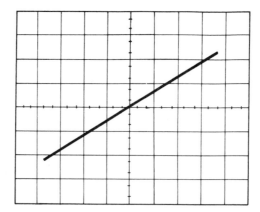

Figure 1-19: Correct oscilloscope reference linearity pattern.

The next step is to set the signal generator frequency to 1 kHz (audio tests used to be made using 400 Hz, but 1 kHz is used more frequently today). At this point, a diagonal line should appear on the scope screen. If the scope amplifiers—both vertical and horizontal—are linear, you'll see a perfectly straight diagonal line across the scope viewing screen, similar to the one shown in Figure 1-19.

If the amplifiers are not linear, you'll see a curved line. This is an indication that the scope needs some adjustment or repair. Refer to the manufacturer's instructions in this case. After you have completed checking the scope amplifiers and are satisfied they will not distort the test signal, proceed as follows.

Procedure:

Step 1. Connect the equipment as shown in Figure 1-20. The load resistor, R, must be capable of dissipating the amplifier output power without changing value, as explained in the section on Power Response Testing. See Sections 1.15 and 1.16.

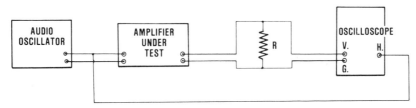

Figure 1-20: Test setup for checking an amplifier's linearity with an oscilloscope.

Step 2. Adjust the signal generator output level until it drives the amplifier to its maximum *undistorted* power out.

Step 3. Finally, check the pattern on the scope. If you see exactly the same pattern that you saw on the reference pattern, the amplifier can be considered to be linear at the frequency being used to excite the amplifier (1 kHz, in this instance), all the way up to the maximum undistorted power. Note that this is not a frequency linearity test.

Sine-Wave Generator and Step Attenuator Method

Test Equipment:

Sine-wave generator, dB reading meter and step attenuator

Test Setup:

See Figure 1-21.

Comments:

The basic requirements are the same for this test as for all other tests; that is, ac line voltage must be stable, all equipment should be left to warm up

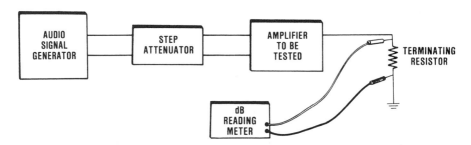

Figure 1-21: Equipment setup for measuring an amplifier's linearity using a step attenuator, audio signal generator and dB reading meter.

at about 1/10 the normal highest operating power, don't overdrive the equipment during measurements, and terminate the amplifier under test with the manufacturer's recommended load impedance, for example, a non-inductive load resistor. The test setup for this linearity measurement is shown in Figure 1-21.

To make a very accurate measurement of the linearity of an audio amplifier, the step attenuator should be capable of being varied in steps of 1/10 dB. This will permit you to determine the exact point when the amplifier departs from linearity. However, for general use, much larger steps may be tolerated.

Procedure:

Step 1. Set the step attenuator for an output signal level of about 20 dB below the amplifier's rated maximum power output (a power ratio of about 100 to 1).

Step 2. Adjust the step attenuator in 1 dB steps (or any other convenient value, if accuracy isn't a major concern), and watch the output signal level meter. For each step down in attenuation, you should see the same amount of increase on the output meter. For example, removing 1 dB of attenuation should produce 1 dB increase on the output meter, if you're removing 1 dB of attenuation at a time.

Step 3. Finally, when the output meter reading starts changing less dBs than the dB changes made on the input signal attenuator, you have reached the amplifier's point of nonlinearity. As you can see, the finer the steps of attenuation, the more accurately you can read the exact departure from linearity.

SECTION 1.12: AUDIO FREQUENCY MEASUREMENTS

Signal Generator and AC Voltmeter Method

Test Equipment:

Good quality ac voltmeter and sine-wave signal generator

Test Setup:

See Figure 1-22.

Comments:

There are numerous ways to measure an audio frequency. Without question, using a digital frequency counter is the easiest and most accurate of all methods. However, it's possible to make a fairly accurate audio frequency measurement using nothing but a good-quality audio signal generator and ac voltmeter. The test setup for the measurement is shown in Figure 1-22.

Procedure:

Step 1. Adjust the output of the signal generator until you have a center scale reading on the ac voltmeter without the unknown signal being connected to the circuit.

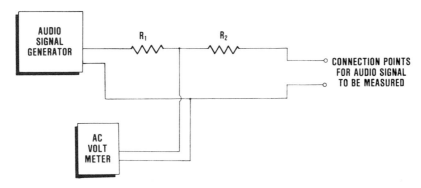

Figure 1-22: Circuit connections for measuring an audio frequency with an ac voltmeter and audio signal generator (See Procedure for R_1 and R_2 values).

Step 2. Next, connect the unknown frequency to the connection points shown in the diagram.

Step 3. Tune the signal generator and watch the ac voltmeter. As you tune the generator, you'll see the ac voltmeter needle start to pulsate. The faster the needle swings back and forth, the farther off frequency you are. The slower, the closer the signal generator is to the unknown frequency.

Step 4. When the generator is set at exactly the same frequency as the unknown frequency, you'll see no movement of the meter needle (zero beat). Finally, read the frequency off the audio signal generator dial. This reading is a quite accurate value of the unknown frequency.

The two resistors (R_1, R_2) shown in Figure 1-22 can be any two resistors of equal value, that you happen to have on hand, which are somewhere near 1,000 ohms (plus or minus a few-hundred ohms). The greatest precision can be gained by setting the standard signal generator output to about ten times the level of the unknown signal amplitude, although lesser amplitudes can be used without too much loss in accuracy. Also, it's good practice to check the standard signal generator to 60 Hz and watch for slow ac meter pulsations. More than likely, there won't be any. However, it's best to check, especially if you can't bring the equipment to zero beat.

Oscilloscope Method

Test Equipment:
   Oscilloscope with a calibrated time base

Test Setup:
   Apply an unknown signal to the vertical input of the scope.

Procedure:
   Step 1. Adjust the oscilloscope sweep rate until you see one motionless cycle on the face of the scope.

   Step 2. Note the width of the cycle by counting the number of divisions it fills on the scope graticule and find the time period by checking the settings of the oscilloscope sweep control.

   Step 3. Now, simply use the formula, frequency = 1/time period. For example, let's say you count 4 graticule divisions for one cycle on the screen. Set the sweep control of the scope at 10 msec per division and the scope multiplier at times 1. In this case, time = 4 × 10 = 40 msec or 0.04 seconds; therefore, dividing this into 1, we get 25 Hz.

Oscilloscope and Calibrated Sine-Wave Generator Method

Test Equipment:
   Oscilloscope and calibrated standard sine-wave generator

Test Setup:
   Apply the standard signal generator output to the external horizonatl input of the scope. Connect the unknown signal to the vertical input of the scope.

Procedure:
   Step 1. Adjust the amplitude of the standard signal generator until it is the same as the unknown signal. You'll know when the amplitudes are equal and the two frequencies are exactly the same (and 90° out of phase) when you see a perfect round circle on the face of the scope.

   Step 2. Read the frequency off the standard signal generator; this is the frequency of the unknown signal. If the unknown frequency is an exact multiple of the standard frequency, 2 times, 3 times, or 4 times as high, you'll see the patterns shown in Figure 1-23 A, B, and C. But, should the unknown frequency be a submultiple, 1/2, 1/3, or 1/4 of the standard, you'll see the patterns shown in Figure 1-23 D, E, and F.

Figure 1-23 shows only a few of the many patterns that are possible to create using different frequency ratios. However, a frequency ratio of 10 to 1 is about the highest ratio you can use because it becomes impossible to count the loops on the scope. For example, if you can count ten loops along the top of B, it indicates that the unknown frequency is ten times as high as the reading on the signal generator you're using for a reference.

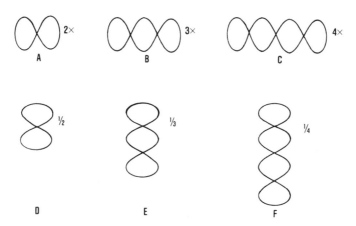

Figure 1-23: Scope patterns and frequency ratios when two sine-wave signals are out of phase. The phase shift is produced automatically as you adjust the frequency.

Dual Trace Oscilloscope and Signal Generator Method

Test Equipment:

Dual trace oscilloscope and signal generator

Test Setup:

Unknown signal to channel A. Reference signal generator to channel B.

Procedure:

Step 1. Adjust the scope for a single steady sine-wave cycle on the channel B trace.

Step 2. Next, adjust the wave pattern on channel A until you see 1, 2, 3, or more, cycles on the screen. If you see only one cycle on each channel, the two frequencies are the same. Two cycles on channel B means the ratio is 2 to 1, three cycles is 3 to 1, and so on. This method is just as accurate as your standard signal generator, which is also true of the preceding method.

SECTION 1.13: SPEAKER IMPEDANCE MEASUREMENT
(SINGLE CONE)

VOM and Audio-Signal Generator Method

Test Equipment:

High impedance VOM, audio-signal generator, and a 20-ohm variable resistor

Test Setup:

See Figure 1-24.

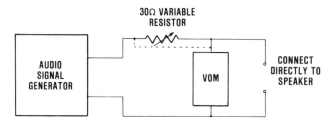

Figure 1-24: Circuit for measuring the impedance of a dynamic speaker.

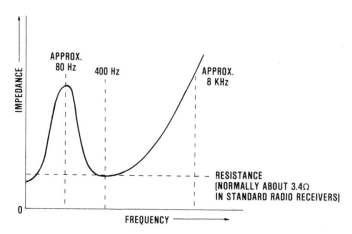

Figure 1-25: Typical single-cone speaker frequency versus impedance plot. It is very apparent that you must use a frequency fairly close to 400 Hz to measure the speaker impedance, or you'll end up with a value much too high.

Comments:

The impedance of a dynamic speaker is usually specified for a certain frequency—typically 400 Hz. You'll find that the dc resistance measured with an ohmmeter is very close to the impedance of the speaker when it is excited with a 400 Hz signal. However, decrease the test frequency and you'll find that the dynamic impedance will increase sharply. On the other hand, increasing the frequency to above 400 Hz will cause the impedance to rise to a high value, but at a lower rate of climb. If you plot a curve of the impedance at different frequencies, it will look like the one shown in Figure 1-25.

Procedure:

Step 1. Measure the dc resistance of the speaker with the VOM.

Step 2. Connect a variable resistor of about twice the value of the dc resistance in series with the speaker. *Note:* Using twice the value will require a resistor of about 10-ohms, for most radio receivers. However, using a 20-ohm resistor will work with almost any speaker impedance. Also, in many of today's high power amplifiers, the watts rating of the resistor may be important if you use the amplifier output as your signal source. In this case, drive the amplifier with the signal generator set at 400 Hz. Other than this change in hookup, use the same setup as shown in Figure 1-24.

Step 3. Set the signal generator to 400 Hz. Connect the VOM to point A and, with the variable resistor, adjust for a convenient voltage reading.

Step 4. Move the VOM to point B and adjust the variable resistor until the voltage reading in Step 3 drops to one-half its previous value. When the voltage at point B exactly equals the voltage at point A, the impedance of the speaker is equal to the value of the resistor.

Step 5. Remove the variable resistor and measure its resistance with your ohmmeter. Your resistance reading is the impedance of the speaker. Incidentally, if you have a resistance box, it can be used in place of the variable resistor to save a few steps. But be sure it can handle the power if the speaker is connected to an amplifier during the measurement.

Oscilloscope and Audio-Signal Generator Method

Test Equipment:

Oscilloscope, audio-signal generator and a 20-ohm variable resistor

Test Setup:

See Figure 1-26.

Procedure:

Step 1. Connect the variable resistor in series with the speaker. Apply a 400 Hz signal to the speaker, through the resistor.

Step 2. Connect the scope vertical input lead to the signal generator's output terminal.

Step 3. Connect the scope horizontal lead to the speaker's ungrounded input terminal.

Step 4. Set the sweep frequency of the scope to view a 400 Hz signal and adjust the vertical and horizontal level controls until you have a line about 3

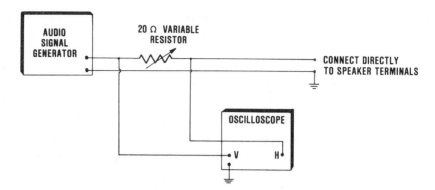

Figure 1-26: Connection for measuring a single-cone speaker impedance with an oscilloscope and audio-signal generator.

or 4 inches long on the viewing screen. The line will probably slant one way or the other, but this is not important. Either way is okay.

Step 5. Adjust the variable resistor and scope controls until you have a line on the viewing screen that slants as close as possible to 45°.

Step 6. When you have the line slanting at 45°, the value of the resistor is equal to the impedance of the speaker. Simply measure the value of the resistor and this value is the speaker impedance at 400 Hz.

SECTION 1.14: SPEAKERS IN PHASE TEST

Test Equipment:
    None

Test Setup:
    See **Procedure.**

Comments:
    Most loudspeakers have coded terminals; sometimes a red dot or a plus sign, and sometimes a red fiber washer under the positive terminal. However, some are not marked. This simple test will work in cases of this type, as well as make a quick check of speaker phasing.

Procedure:
    Step 1. Connect the two speakers to the stereo amplifier output terminals.

    Step 2. Reverse the speaker hookup.

    Step 3. The speaker connection that produces the best bass response is the better connection. In other words, the speakers are operating in phase.

SECTION 1.15: POWER RESPONSE TEST—LOW AND
MEDIUM POWER AMPLIFIERS

Typical low-cost audio power meters provide constant indication of amplifier output power and will read up to 100 or 200 watts rms. Generally, these instruments have two (for stereo) root-mean-square (rms) meters that measure the average power output in one or two ranges. For example, the 0-2 and 0-200 watt ranges. Some have LEDs to track short bursts of music energy and to help spot distortion.

These instruments are easy to use and have simple "hookups." However, there are times when more exacting information is needed. This section and the next contain procedures that can be used in these cases.

Test Equipment:

Sine-wave generator, scope, ac voltmeter and terminating resistor

Test Setup:

See Figure 1-27.

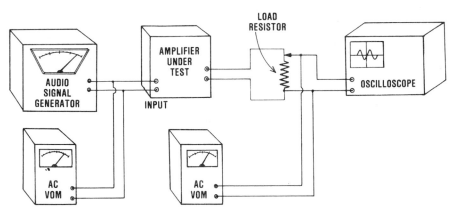

Figure 1-27: Test equipment setup for power response testing. *Note:* Although two meters are shown, it should be understood that only one meter is necessary if the signal generator has a metered output.

Comments:

Power response is defined as the frequency response capabilities of an amplifier running at or near its full rated power. There is a standard test procedure used by many service centers and manufacturers to measure an amplifier's power response. The measurement is made with the amplifier connected to a resistor with a very low reactive component (preferably no more than 10 percent). The resistor value is the same as the amplifier output impedance and must be capable of dissipating the full load of the amplifier while maintaining its resistance value very close to its rated value (ideally, within plus or minus 2 percent).

For professional jobs, all test equipment should be of very good quality. For example, the signal generator distortion shoulld not exceed 20 percent of the measured distortion of the amplifier being tested and should be capable of maintaining the test frequencies within plus or minus 2 percent.

Procedure:

Step 1. Connect the signal generator to the amplifier input and set it to a 1 kHz test frequency.

Step 2. The scope and ac voltmeter are connected across the load resistor, as shown.

Step 3. Set the volume control to maximum and follow the manufacturer's specs for all other controls. Most of the time the other controls—for example, tone controls—are set to the center position, except for special circuits such as bass boost, which is usually set to off.

Step 4. Apply power and adjust the signal generator until you see a sine-wave pattern on your scope.

Step 5. Watch the waveform on the scope as you keep increasing the input signal level (probably, you'll have to reduce the scope vertical gain to keep the pattern on the screen). When you see the first signs of distortion on the sine wave (you'll see flattening peaks), reduce the signal generator output level until the clipping of the peaks just barely disappears. This is the amplifier's maximum undistorted power output level.

Step 6. Now we come to the reason for the two ac voltmeters, which is two-fold. One, it is necessary to hold the input voltage constant over the frequencies you decide to check (this isn't a problem with a calibrated attenuator, if the AF generator has one), and two, to compute the amplifier power out, you'll need to know the rms voltage developed across the load resistor. To do this, simply use the formula, $P = E^2/R$, where P is watts, E is the voltage measured across the load resistor and R is the value of the load resistor.

SECTION 1.16: POWER RESPONSE TEST— HIGH POWER AMPLIFIERS

Test Equipment:
Sine-wave generator, four 8-ohm load resistors, and a high input impedance ac voltmeter

Test Setup:
See Figure 1-28.

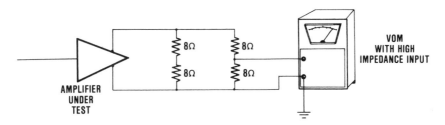

Figure 1-28: Voltage divider network that can be used to measure the output of a high power audio amplifier.

Comments:

Not long ago, an audio amplifier was considered to be a high power amplifier if it could produce 50 watts of continuous power when driving an 8-ohm speaker. Today, there are amplifiers that generate 100 watts or more. Therefore, sometimes it's necessary to measure power output using a voltage divider network. A simple voltage divider load resistor circuit is shown in Figure 1-28.

Procedure:

Step 1. Let's assume the amplifier is rated at 100 watts of output power and has an 8-ohm output impedance. To make the power measurement, connect four 8-ohm resistors, as shown in Figure 1-28.

Step 2. Connect your VOM across one of the 8-ohm resistors, as shown. In this case, you should read approximately 1.767 volts rms on your ac voltmeter.

Step 3. Now, using the formula, $P = E^2/R$, we get 25 watts. However, the measurement was taken across the voltage divider so we'll have to multiply by 4 to get the true power reading. The end result is 100 watts, which indicates that the amplifier is performing according to specs.

SECTION 1.17: INPUT SIGNAL LEVEL TEST

Test Equipment:

See Figure 1-27.

Comments:

This test uses exactly the same setup that was explained for Power Response Testing (See Figure 1-27); however, you don't need the voltmeter on the output circuit.

Procedure:

Step 1. Adjust your signal generator for maximum undistorted power, as explained in the section on Power Response Testing (Section 1.15).

Step 2. Next, read the voltage across the amplifier input circuit, or take the reading from the signal generator's calibrated attenuator. As an example, if you can produce about 400 millivolts on the ac voltmeter without causing clipping of the sine-wave output signal you see on your scope, you can consider the amplifier to be in good working order. This test is called a sensitivity test by some manufacturers, and its purpose is to check the amount of input signal needed to drive the amplifier.

SECTION 1.18: DC POWER SUPPLY VOLTAGE TEST
AND PROBABLE TROUBLE

Test Equipment:
 VOM

Test Setup:
 None

Comments:
 As with most tests, start out by checking the power supply's primary voltage supply; i.e., the ac power line.

Procedure:
 Step 1. Measure the power supply output voltage with the load disconnected. Refer to Table 1-1 if you have other than a normal reading.

CIRCUIT	LOW DC VOLTAGE OUTPUT	NO DC VOLTAGER OUTPUT
HALF-WAVE	INPUT FILTER CAPACITOR OPEN	DIODE OPEN, SURGE RESISTOR OPEN, INPUT CAPACITOR SHORTED
FULL-WAVE	INPUT FILTER CAPACITOR OPEN	DIODES OPEN OR INPUT FILTER SHORTED
HALF-WAVE DOUBLER	INPUT FILTER CAPACITOR OPEN	OPEN INPUT DOUBLER CAPACITOR, DIODE OR SURGE RESISTOR
FULL-WAVE DOUBLER	INPUT FILTER OPEN, ONE DIODE OPEN; ONE DOUBLER OPEN	FILTER CAPACITOR SHORTED
FULL-WAVE BRIDGE	INPUT FILTER CAPACITOR OPEN NOTE: ONE OPEN DIODE IN BRIDGE WILL NOT REDUCE DC VOLTAGE BY VERY MUCH, BUT WILL INCREASE RIPPLE AMPLITUDE	FILTER CAPACITOR SHORTED

Table 1-1: DC power supply testing and troubleshooting chart.

Step 2. If the output voltage of the power supply is correct, connect the normal load to the supply.

Step 3. Measure the power supply output voltage. If it is up to the value shown in the schematic, fine. If it is below normal, you may have an overload somewhere in the load.

SECTION 1.19: MEASURING THE EFFECTIVE LOAD
IMPEDANCE OF A DC POWER SUPPLY

Test Equipment:
   Voltmeter and ammeter

Test Setup:
   Connect the dc power supply to the load under consideration with an ammeter, all in series. Connect a dc voltmeter across the power supply output terminals.

Procedure:
   Step 1. Measure the power supply output voltage with the load connected and operating.
   Step 2. Measure the power supply output current with the load connected and operating.
   Step 3. Calculate the effective load impedance with the formula $R = E/I$; for example, with a voltmeter reading of 12 volts and a current reading of 2 amperes, $R = E/I = 12/2 = 6$ ohms.

SECTION 1.20: LINE VOLTAGE REGULATION
MEASUREMENT—VOLTAGE REGULATED POWER SUPPLY

Test Equipment:
   High input impedance voltmeter, variable load resistor, variable voltage line transformer (Variac) and ammeter

Test Setup:
   See Figure 1-29.

Comments:
   Line voltage regulation frequently is measured under constant load current conditions (normally at one-half full-load current). However, better results will be obtained by measuring the regulation at both zero load and full load because, in most cases, you'll find that the power supply performs better at zero load than at full load, and this fact should be taken into consideration.

Procedure:
   Step 1. Adjust the Variac for a power supply input line voltage according to the power supply manufacturer's highest operating voltage specification. If

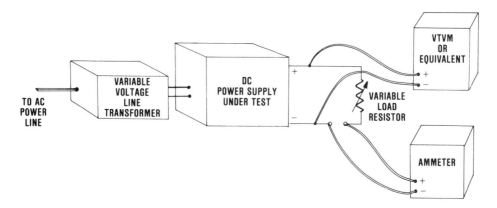

Figure 1-29: Test setup for measuring a voltage regulated power supply line voltage regulation.

you have no specs, set it to 130 volts. Measure the dc voltage output of the power supply and note the current reading.

Step 2. Adjust the Variac for a power supply input line voltage according to the manufacturer's lowest operating voltage specification. If you don't have the specs, set it to 100 volts. Measure the dc voltage on the power supply output when the current is the same as you had in Step 1. *Adjust the load resistor to maintain a constant current reading.*

Step 3. Calculate the voltage that is half-way between the high and low settings of the Variac voltage input to the power supply. Set the Variac to this voltage and measure the power supply dc output voltage. This will be the so-called, *mid-scale voltage* in the next step.

Step 4. Calculate the voltage regulation percentage using the formula:

line voltage regulation (%) = [highest dc voltage − lowest dc voltage/dc output voltage(Variac set at mid-scale ac voltage input)] × 100

SECTION 1.21: CHECKING A POWER SUPPLY'S RIPPLE AND OTHER PERIODIC AND RANDOM DEVIATIONS

Test Equipment:
Oscilloscope and load resistor selected to operate with a full load on the power supply under test. See **Comments** for operating conditions during test.

Test Setup:
See **Procedure.**

Comments:

Transistors and other solid state devices do not like transient spikes and surges in their power supplies. In fact, rectification spikes can destroy the logic elements in some digital systems. Sometimes you'll find all periodic and random deviations that show up on the output of a power supply, grouped together under the heading of "PARD." This stands for Periodic and Random Deviation, and the tests normally are made at 50 percent load current. The rectifier switching spikes will be seen on a scope as periodic vertical lines. Another deviation might be shock-excited oscillations. These will be seen on a scope as a "ringing" sine wave. Ripple usually is measured under full load power and will be seen on a scope as a sine wave (possibly, as a sawtooth), having a frequency of 60 Hz to 120 Hz.

Procedure:

Step 1. Connect the scope's vertical input probe across any of the filter capacitors except the input capacitor. You'll always find some ripple voltage on the input capacitor—probably about 10 volts peak-to-peak in transistor equipment power supplies.

Step 2. Adjust the scope vertical gain to maximum and set the scope to the ac mode of operation.

Step 3. Place a load resistor that will provide the desired load, on the power supply and energize the system.

Step 4. Look at the scope's viewing screen. If you see any ac signal, the power supply is under suspicion. All you want to see is a simple dc line. For example, ripple levels in well-filtered power supplies should be very low. Even 0.1 volts peak-to-peak sometimes is too much.

SECTION 1.22: HOW TO CONVERT RIPPLE VOLTAGE TO dBs BELOW THE MAXIMUM DC OUTPUT VOLTAGE

Test Equipment:

Voltmeter and/or oscilloscope

Test Setup:

See **Procedure.**

Comments:

Ideally , an unregulated dc power supply should have a ripple voltage about 80 to 100 dB below the dc voltage under full load, when operating at its rated voltage output. A regulated supply should be better—about 120 dB below the output voltage. However, these values are much better then you'll find in the average supply.

Procedure:

Step 1. Measure the maximum dc voltage output under full load. For best results, the load should be a non-reactive resistor. However, any load with little reactance, such as an amplifier, etc., will work if accuracy is not a critical factor.

Step 2. Measure the ripple voltage with an oscilloscope or other suitable instrument, such as a good quality ac voltmeter.

Step 3. Calculate the ripple voltage level below the maximum dc level using the formula:

$$dB = 20 \log_{10} \text{(dc output voltage)/(ripple voltage)}$$

For example, a regulated supply has 40 volts dc on its output and its ripple voltage is measured and found to be 150 microvolts rms. In this case,

$$dB = 20 \log (40/150 \ \mu V) = \text{approximately } 108$$

SECTION 1.23: HOW TO CONVERT POWER OUT FROM VUs TO WATTS

It isn't unusual to find that manufacturer's specs give an amplifier output in volume units (VUs), but frequently it's necessary to convert VUs to watts. An example of how to do this can best be shown with a practical in-shop measurement.

Let's assume your VU meter reads 30 VU and you have exactly 0 VU (1 milliwatt in 600-ohms) input signal. Now, all you have to do is use the following formula and do your work as shown.

$$\text{output power} = \text{antilog VU/10} = \text{antilog 30/10} = \text{antilog } 3 = 1{,}000$$

Your answer will be in the same units as your reference level; milliwatts in this case. So the answer is 1,000 milliwatts, or simply 1 watt.

SECTION 1.24: HOW TO CONVERT VOLTAGE READINGS TO dBs

Experience is really the best teacher of dB/volt conversion methods. But it won't be long, if you follow these tips, before you will be able to convert the readings, even with amplifiers that have unequal input and output impedances. Most electronics technicians are familiar with calculating voltage gain in dBs for an amplifier that has equal input and output impedances. However, to review quickly, it requires two ac voltage measurements; 1) the voltage on the amplifier input E_1; and 2) the voltage across the output load termination E_2. Then simply use the formula, $dB = 20 \log_{10} (E_2/E_1)$. Table 1.2 shows voltage and current ratios converted to dBs. The decible calculations have been rounded off slightly, but they are sufficiently accurate for all the work most of us will encounter.

If the amplifier has unequal input and output impedances and you want to calculate the voltage gain in dBs, you have to use the formula:

$$dB = 20 \log_{10} (E_2 \sqrt{Z_1} / E_1 \sqrt{Z_2})$$

To find E_1 and E_2, measure the voltage at the amplifier input (E_1) and measure the voltage across the output load termination (E_2). Next, the value for Z_1 is the amplifier input impedance and Z_2 is the output load impedance.

Impedance depends on the frequency you use during the measurement, because both X_L and X_C vary with frequency. Therefore, any impedance value you come up with is valid only for that frequency at which it was calculated or measured. Typically, 1,000 Hz is used as a reference frequency for measurements between 20 and 20,000 Hz.

VOLTAGE AND CURRENT RATIO	DECIBELS	VOLTAGE AND CURRENT RATIO	DECIBELS	VOLTAGE AND CURRENT RATIO	DECIBELS
1.0116	0.1	1.4962	3.5	12.589	22.0
1.0233	0.2	1.5849	4.0	15.849	24.0
1.0351	0.3	1.6788	4.5	19.953	26.0
1.0471	0.4	1.7783	5.0	25.119	28.0
1.0593	0.5	1.8836	5.5	31.623	30.0
1.0715	0.6	1.9953	6.0	39.811	32.0
1.0839	0.7	2.2387	7.0	50.119	34.0
1.0956	0.8	2.5119	8.0	63.096	36.0
1.1092	0.9	2.8184	9.0	79.433	38.0
1.1220	1.0	3.1623	10.0	100.000	40.0
1.1482	1.2	3.5481	11.0	125.89	42.0
1.1749	1.4	3.9811	12.0	158.49	44.0
1.2023	1.6	4.4668	13.0	199.53	46.0
1.2303	1.8	5.0119	14.0	251.19	48.0
1.2589	2.0	5.6234	15.0	316.23	50.0
1.2882	2.2	6.3096	16.0	398.11	52.0
1.3183	2.4	7.0795	17.0	501.19	54.0
1.3490	2.6	7.9433	18.0	630.96	56.0
1.3804	2.8	8.9125	19.0	794.33	58.0
1.4125	3.0	10.0000	20.0	1000.00	60.0

Table 1-2: Voltage to current ratios converted to decibels, and vice versa.

SECTION 2

Practical Guide to Maintaining Digital Equipment

SECTION 2.1: A PRACTICAL INTRODUCTION TO DIGITAL ICs

Transistor-transistor logic refers to the type of electronic components used to construct certain integrated circuits. This type of logic is manufactured using multiple-emitter transistors in the input circuit, and the output transistors are switching transistors, which greatly reduces matching problems. See Figure 2-1 for a schematic of a TTL IC NAND gate.

The 7400 series is currently inexpensive and quite popular. This is mostly due to the low cost (a few cents) and to the fact that it has a comparatively high operating speed, fairly low power dissipation, large multiple load capabilities (fan-out), and excellent electrical noise margin. Incidentally, the example TTL gate shown in Figure 2-1 has only two emitters on input transistor Q1. This is not always the case. A four-input TTL gate will have four emitters of a single transistor as the input element. On the other hand, a 7400 series hex inverter (the 7405) has only one emitter on the input transistor. The main point is that TTL circuits use bipolar transistors in their design.

Compared to bipolar types, metal-oxide semiconductor (MOS) devices offer the advantage of low power requirements, high input impedance, and high gain, but these advantages are offset for many applications by a limited power capability, and they are easily damaged by sudden changes in voltage (voltage transients) and short-term overloads.

An area of MOS technology that uses P-channel and N-channel metal oxide field effect transistors (MOSFETs), fabricated on the same chip (integrated circuit before packaging) in a complementary switching arrangement, is called CMOS. An illustration of a typical CMOS circuit configuration is shown in Figure 2-2.

The CMOS logic family uses voltage-sensitive MOSFETs as opposed to current-sensitive bipolar transistors used in the TTL family of ICs, resulting in an ultra-low power requirement. To see why CMOS requires so little power,

36

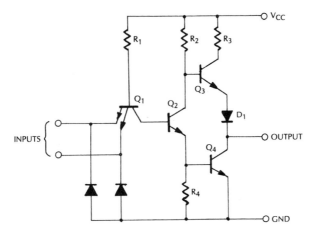

Figure 2-1: Transistor-transistor logic NAND gate. Note the multiple-emitter transistor on the input and the switching transistors on the output. This type of construction is called TTL logic.

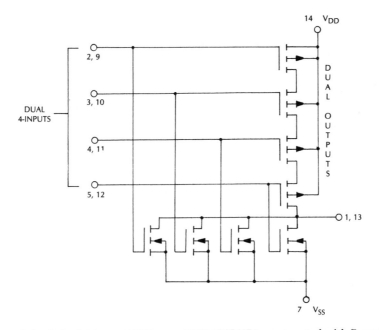

Figure 2-2: A dual 4-input NOR gate (MC14002UB) constructed with P- and N-channel enhancement mode devices in a single monolithic structure (complementary MOS). This circuit schematic shows dual inputs (pins 2, 3, 4, and 9, 10, 11, 12) and dual outputs (pins 1 and 13): i.e., the schematic is one-half the IC but shows all pin numbers.

let's examine one single CMOS gate extracted from the dual 4-input NOR gates shown in Figure 2-2. The gate (an inverter) we are going to examine is shown in Figure 2-3, with pin numbers of the MC14002UB for easy identification.

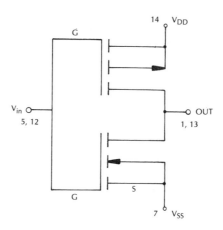

Figure 2-3: Basic CMOS gate (an inverter) extracted from the schematic of the dual 4-input NOR gate shown in Figure 2-2.

First, notice that a complementary (the "C" in CMOS) pair of enhancement-mode MOSFETs comprises this inverter. Although the rest of the CMOS logic elements appear more complex, they all are complementary MOSFETs. The MC14002 ICs are constructed with P- and N-channel enhancement mode devices, as we have said. A gate such as this is normally off, displaying a high resistance between drain and source. To turn it on, you must have a sufficiently large voltage between the gate and source. In this circuit, there are two MOSFETs: the upper one is a P-channel MOSFET, and the lower is an N-channel device. The two form a series circuit between V_{DD} (pin 14) and V_{ss} (pin 7, usually ground).

If the voltage on pin 5 (input) is low, the upper MOSFET is on and the lower MOSFET is off. In other words, because this is a series circuit—between V_{DD} and V_{ss} in this case—an opposition to current flow is offered in either state. But during the input transition (input going toward V_{DD} or V_{ss}), some current will flow through the series circuit composed of the two MOSFETs. That's why CMOS requires so little current; no change in input level, no supply current drawn.

When servicing solid state digital devices, you may encounter *emitter-coupled logic* (ECL). As in the standard TTL family, emitter-coupled logic uses transistors in its design, but ECL is a form of digital logic that eliminates transistor storage time as a speed-limiting characteristic, permitting very high speed operation. As you would probably assume, "emitter-coupled" refers to the manner in which transistor amplifiers (differential amplifiers) are electronically coupled within the ICs.

The differential input amplifiers of this logic family provide high-impedance inputs and voltage gain within the circuit. By referring to Figure 2-4, you'll notice that the output of this basic ECL gate is an emitter follower. This circuit restores the logic levels and provides low output impedance for good line driving and multiple load (fan-out) capability.

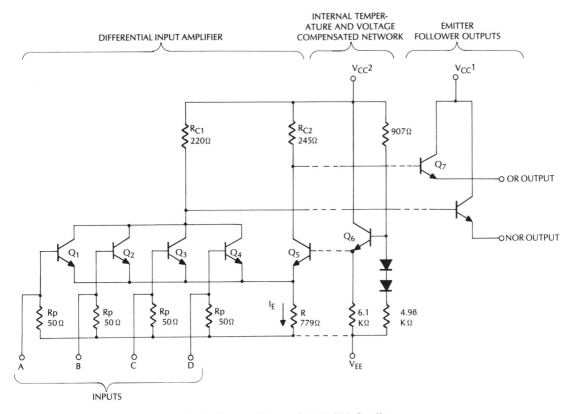

Figure 2-4: A MECL basic gate (Motorola's 10,000 family).

Motorola has offered MECL circuits in four logic families: MECL I, MECL II, MECL III, and MECL 10,000. The MECL III logic family is listed as the fastest standard logic available, designed to provide the high speed required by computers, communications, and instrument systems.

It has been pointed out that the best solution to the TTL power requirement problem is to use CMOS ICs instead. However, technicians must be ever-conscious of whether they are working with TTL, CMOS, or other logic families because these are classifications of fabrication techniques, as has been explained. These procedures for manufacturing an IC determine critical electrical characteristics such as operating voltage levels, input and output impedance (important when interfacing ICs), drive capabilities (fan-in and fan-out) and, of course, operating speed.

CMOS devices offer a wider range of operating voltage levels and higher input impedances. These ICs are an attractive option for relatively low speed circuits that are to operate from standard battery power sources of 6V, 9V, and 12V. At this point you may be wondering, "Why not combine CMOS and TTL ICs?" You'll find that, because of interfacing problems, using both these devices in a single circuit isn't easy. For example, a TTL quad-2 input NAND gate (N7403) has a normalized fan-out from a single gate of a maximum of ten TTL loads, whereas a CMOS quad-2 input NAND gate (MC1411B) has only a maximum capability of driving two low-power TTL loads. Also, the N7403's recommended operating voltage (V_{CC}) is a minimum of 4.5V to a maximum of 5.5V. The MC14011B has a recommended voltage (V_{DD}) of -0.5 to $+18$Vdc (reference to V_{SS}).

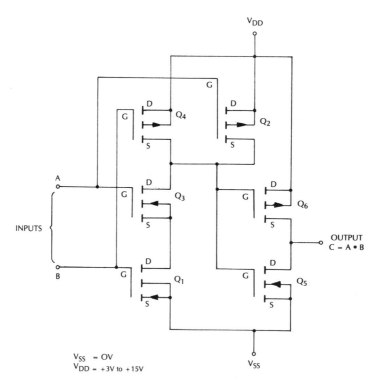

Figure 2-5: Basic CMOS 2-input AND gate. Typically, such a gate will be found in an IC such as a MC14081, a quad-2-input AND gate (a pin-for-pin replacement for the corresponding CD4000 series).

Sometimes you may have trouble with logic level mismatching. In fact, in many cases, every time you interface between the two different families it may require some special circuitry, even if you are only driving a low-power TTL with a CMOS. What this all means is that you should take care when servicing both CMOS and TTL in the same system.

Today, most technicians are familiar with TTL family AND, OR, and NAND gates. However, if you are not, see Section 2.4. Also, many technicians are not at ease with the newer CMOS gates. Figure 2-5 shows a basic CMOS AND gate circuit.

To see how this circuit works, suppose you apply a logic low (a near V_{SS} potential) to input B (gate electrode of Q3). At this point, the series circuit composed of Q1 and Q3 will be turned off. On the other hand, while it is turning off Q1 and Q3 (N-channel MOSFETs), it is turning on Q4, a P-channel MOSFET. Next, notice Q4 and Q2 are in parallel. This means that these two transistors are both acting as a closed switch (conducting). That, in turn, causes the gates of the outputs (Q5 and Q6) to be at a potential very near V_{DD}. It turns P-channel FET (Q6) off, and the N-channel (Q5) on. The

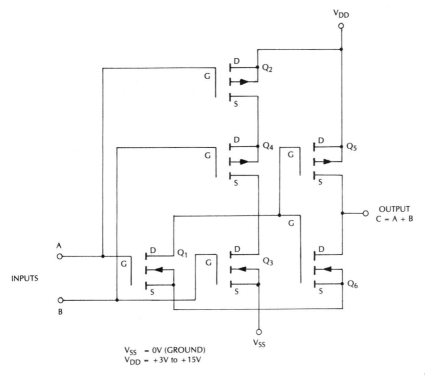

V_{SS} = 0V (GROUND)
V_{DD} = +3V to +15V

Figure 2-6: Basic CMOS 2-input OR gate. A gate such as this would be the internal structure of an IC such as the MC1407B, a quad-2-input OR gate.

output is now near V_{DD} (a logic high). The result: input A and input B arrive at the output C as input A and B, or $C = A \cdot B$. Figure 2-6 is the schematic of a single CMOS OR gate circuit. Typically, four of these gates would be found in an IC such as a MC1407B, a quad-2-input OR gate.

To get some idea of the function of the CMOS OR gate circuit, let's assume that you place a logic high (V_{DD}) on point A. Notice that this applies V_{DD} to both Q1 and Q2 gates. Q1 is an N-channel FET, so it switches on. Q2 is a P-channel FET, therefore it switches off. Looking at the schematic very closely, you'll see that Q1 and Q3 are in a parallel circuit and Q2 is part of a series circuit that includes Q4. What this means is that when Q1 is on, the gates of output FET's Q5 and Q6 will drop down to the V_{SS} potential, and turning off Q2 cuts the gates of those output FETs from V_{DD}. The output FETs now have a potential on these gates of V_{SS}, which means Q5 is on and Q6 is off. The output (C) is now very near V_{DD}.

If we go through the circuit again with a logic high (V_{DD}) applied to the input B, it will turn out to be exactly the same; i.e., the output C will be at a logic high. Furthermore, if you apply a logic low (V_{SS}) to the input (A or B), you will produce a logic low on the output. In other words, if you connect both inputs to V_{SS}, it should make your IC output go to a logic low or some voltage near V_{SS}. Connect both inputs to V_{DD} and the output should be at a logic high (V_{DD} or some potential near V_{DD}). This means that the circuit performs as an OR gate.

Although other CMOS logic elements are more complex, they all use complementary MOSFETs, and the preceding circuit description captures the most essential characteristics of the CMOS logic family. However, CMOS does have an important rival, the NMOS, which is a faster-operating logic family. In fact, CMOS is only a fourth as fast as TTL. But this problem is beginning to disappear with a new generation of CMOS chips. Both Motorola and National Semiconductor have introduced a family of CMOS chips that are functionally similar to standard TTL parts. This family is designated 74HCxx. Other logic elements such as Schottky TTL will be discussed in a later section.

SECTION 2.2: CODING TECHNIQUES USED
IN DIGITAL SYSTEMS

As you may know, any data used by digital ICs must be represented by binary notation; i.e., 1s and 0s. Also, as you probably have guessed by now, before a base 10 number can be used by a digital circuit, it must be converted to high and low logic voltage levels.

Let's now look a bit deeper into the various numbering systems used in digital circuits (for example, base 2—binary: base 8—octal; and base 16—hexadecimal) and see why one or the other is more convenient to use. First

DECIMAL	OCTAL	BINARY
0	0	000
1	1	001
2	2	010
3	3	011
4	4	100
5	5	101
6	6	110
7	7	111

Table 2-1: Relationship of binary based numbers to each octal based number and its equivalent base 10 number.

refer to Table 2-1, which illustrates the relationship of each binary number and each octal number to the equivalent base 10 number.

If you don't already know, you'll be surprised at what the octal numbering system can do for you. Notice that each octal based number has a binary equivalent that has three digits (1s and 0s). The important point is that a binary number can be divided into groups of three digits each, starting from the right-hand side of the binary number, and each group of three binary digits may be represented by its octal equivalent. In fact, any binary number can be converted directly from an octal number, or any octal number can be written directly from a binary number. For example:

$$\begin{array}{cc} 010 & 101 \text{ binary} \\ 2 & 5 \text{ octal} \end{array}$$

Let's take another example of this technique and carry it a bit farther. Given a binary number containing many 1s and 0s, convert it to a decimal number using the octal system. For instance, what is the decimal equivalent of the binary number 1101010100?

$$\begin{array}{cccc} 1 & 101 & 010 & 100 \text{ binary} \\ 1 & 5 & 2 & 4 \text{ octal} \end{array}$$

Reading right to left:

$$\begin{array}{rcl} 4 \times 8^0 & = & 4 \\ 2 \times 8^1 & = & 16 \\ 5 \times 8^2 & = & 320 \\ 1 \times 8^3 & = & \underline{512} \\ & & 852 \text{ decimal} \end{array}$$

The conclusion to be drawn is that the octal numbering system can provide a shorthand method for bridging the gap between decimal and binary. Although digital circuits rarely use octal directly, the octal numbering system is encountered when working with various ICs.

Now that we have discussed the octal system, let's examine the hexadecimal (base 16). As you have seen, the octal number system can be used to simplify binary-to-decimal conversion. In this case, we have shown that three binary digits are represented by one octal digit. The hexadecimal system (from here on out we will call it simply the "hex system") makes conversion even easier because each hex digit represents four binary digits. What this means is that 16 binary digits can be represented with six base eight symbols or four hex symbols. Obviously, four hex symbols are easier to work with. However, as we know, the decimal system has only ten digits, or characters, (0,1,...,9), and, of course, six additional characters are required for the hex system. In this system the letters A, B, C, D, E, and F are used, in addition to the numbers 0 through 9. Table 2-2 shows the relationship between decimal, binary, and hex.

To illustrate the usefulness of the hex system, we will convert a 16-digit binary number to its base 10 equivalent. For our example, we will use binary number 1101111101010100. The first step is to divide the binary number into groups of four each, starting from the right and then write the hex equivalent

DECIMAL	BINARY	HEX
0	0	0
1	1	1
2	10	2
3	11	3
4	100	4
5	101	5
6	110	6
7	111	7
8	1000	8
9	1001	9
10	1010	A
11	1011	B
12	1100	C
13	1101	D
14	1110	E
15	1111	F
16	10000	10
17	10001	11
18	10010	12
25	11001	19
26	11010	1A
27	11011	1B
32	100000	20
33	100001	21

Table 2-2: Relationship between decimal, binary, and hex numbers.

of each group (use Table 2-2 to find the hex equivalent). Your work should look like this:

1101	1111	0101	0100	binary
D	F	5	4	hex

Now we know that DF54 is the hex equivalent of the binary number 1101111101010100. Our next step is to convert the hex DF54 to decimal, like this:

D F 5 4

$$
\begin{aligned}
4 \times 16^0 &= 4 \\
5 \times 16^1 &= 80 \\
(F = 15) \times 16^2 &= 3840 \\
(D = 13) \times 16^3 &= \underline{53248} \\
&\text{decimal } 57172
\end{aligned}
$$

A term you will certainly run across sooner or later in your daily work is American Standard Code for Information Interchanges. You will probably see this written in most of today's service literature as ASCII (usually pronounced *askey*). Therefore, let's examine this alpha-numeric code system.

The ASCII Alphanumeric Code System

To begin, *machine language* (another term for the binary high and low logic 1s and 0s we previously discussed) is the *only* language digital equipment can work with; i.e., understands or recognizes. The ASCII standard code is used for interchange of information (whether input or output) between *peripherals* such as keyboards and line printers. For example, some computer systems have an ASCII keyboard with a numeric keypad.

ASCII is an 8-bit code, and therefore is excellent for hex representation. Incidentally, you will find that only seven digits are actually required for ASCII. However, it is usually considered to be an 8-bit code. Now, since hex-to-binary conversion is relatively simple, as we have shown, and is used in integrated circuits, ASCII can be adapted to any personal computer system; in fact, to any instrument using a keyboard. Table 2-3 shows some of the ASCII Codes.

Table 2-4 shows the conversion between ASCII and hex. To make a conversion, select a letter, symbol, or number from the ASCII character column shown. We will use the dollar sign ($) for our example. Next, locate the dollar sign on the ASCII-to-hex conversion table. Then read up the column (up to the number 2). This is your most significant hex digit (MSD). Now, move horizontally to the left and find the least significant hex digit (4). Thus, hex 24 equals the dollar sign in ASCII, or $ equals 24 in hex, whichever way you would like to look at it. This process can be reversed. Table 2-4 shows that hex 24 equals $, and Table 2-3 shows that ASCII code $ equals binary 0100100.

CHARACTER	ASCII CODE	CHARACTER	ASCII CODE
@	1000000	FORM FEED	0001100
A	1000001	CARRIAGE RETURN	0001101
B	1000010	RUBOUT	1111111
C	1000011	SPACE	0100000
D	1000100	!	0100001
E	1000101	,,	0100010
F	1000110	#	0100011
G	1000111	$	0100100
H	1001000	%	0100101
I	1001001	&	0100110
J	1001010	,	0100111
K	1001011	(	0101000
L	1001100	)	0101001
M	1001101	*	0101010
N	1001110	+	0101011
O	1001111	'	0101100
P	1010000	—	0101101
Q	1010001	.	0101110
R	1010010	/	0101111
S	1010011	0	0110000
T	1010100	1	0110001
U	1010101	2	0110010
V	1010110	3	0110011
W	1010111	4	0110100
X	1011000	5	0110101
Y	1011001	6	0110110
Z	1011010	7	0110111
[	1011011	8	0111000
\	1011100	9	0111001
]	1011101	:	0111010
↑	1011110	;	0111011
NULL	0000000	<	0111100
HORIZ TAB	0001001	=	0111101
LINE FEED	0001010	>	0111110
VERT TAB	0001011	?	0111111

Table 2-3: ASCII code.

It is important to keep in mind that units such as microprocessors *do not* use ASCII. This code must be converted to binary for use in a microprocessor and converted back into ASCII for use at the keyboard, etc.

SECTION 2.3: TRANSMISSION OF DIGITAL DATA

The transfer of digital data between a peripheral and a typical micro-computer (or similar system) must utilize circuits to convert the data instructions into bytes that are compatible with the system's internal circuits.

		MOST SIGNIFICANT HEX DIGIT							
		0	1	2	3	4	5	6	7
LEAST SIGNIFICANT HEX DIGIT	0	NUL	DLE	SP	0	0	P	\	p
	1	SOH	DC1	!	1	A	Q	a	q
	2	STX	DC2	"	2	B	R	b	r
	3	ETX	DC3	#	3	C	S	c	s
	4	EQT	DC4	$	4	D	T	d	t
	5	ENQ	NAK	%	5	E	U	e	u
	6	ACK	'SYN	&	6	F	V	f	v
	7	BEL	ETB	/	7	G	W	g	w
	8	BS	CAN	(	8	H	X	h	x
	9	HT	EM	)	9	I	Y	i	y
	A	LF	SUB	*		J	Z	j	z
	B	VT	ESC	+		K	[	k	‹
	C	FF	FS	,	<	L	\	l	:
	D	CR	GS	–		M	]	m	‡
	E	SO	RS	.	>	N	↑	n	~
	F	SI	US	/		O	←	o	DEL

Note: Parity bit in most significant hex digit not
included and characters in columms o and
1(as well as SP and DEL) are non-printing.

Table 2-4: ASCII-to-hex conversion table.

Data bytes—for example, 00011100, one data byte (hex 1C)—are transmitted in both parallel and serial form from the peripheral to the computer. However, the problem with parallel transmission (although it is the fastest) is that, typically, you will need at least eight lines (or eight-line cable) plus other lines for such things as control flag, service request, interrupt lines, etc.

Frequently, parallel transmission cables have 40 lines. But, many computer systems transmit their data bytes over a single telephone line (or a pair of lines) and use a special serial I/O IC, often called an *asynchronous I/O.* This input/output device is identified as "RS-232-C."

Asynchronous transmission of data in serial form does not require a synchronizing clock (a transmission technique called *synchronous* does require one) to be transmitted with the data and one character need not closely follow another. This means that blank spaces of various lengths can be present between the single characters being transmitted.

The binary code (generally from 5 to 8 bits and often called a "character") which comprises a data character, and synchronizing *start* and *stop* elements, is shown for asynchronous transmission in Figure 2-7.

The binary bits and start/stop elements must be transmitted at a specified rate. This rate is called *baud rate* as most of us know. However, in case you don't, baud is a unit of transmission speed derived from the duration of the shortest signal element. Speed in bauds is the number of code elements transmitted per second and normally one element is one data bit interval.

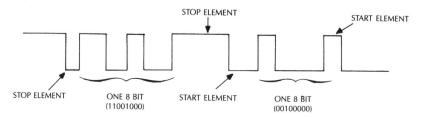

Figure 2-7: Illustration of asynchronous data transmission.

The baud rate is equal to the bit rate if one stop bit is used. But for systems that use more than one stop bit, you cannot determine the baud rate by referring to the bit rate. That is, the baud rate does not equal the bit rate in these systems.

When you turn on or reset some computers, you will see a prompt on the screen in question form (CASS?). In this case, the computer is asking what baud rate you want to use. You must type in H or L, for high or low. Typically, a low baud rate is 500 and a high is 1500. Actually, 500 baud is about 63 characters per second and 1500 baud is around 190 characters per second.

It should be pointed out that if you are using a program that is written for a certain baud rate (for instance, 300), you must use the same baud rate at both ends of the transmission line. For example, from one computer to another, both computers must send and receive at 300 baud (in this instance). However, on an initial setup between two MPUs of the same type, you can use several different baud rates. Typical values that may be used are 110, 150, 300, 600, 1200, 2400, 4800, 9600, 19200 (although at this last rate, your error can be over 3 percent). If your computer does not match the requirements of the device you are using, you must change the baud rate of your computer. *Note:* There are other characteristics that may have to be matched such as word length, parity, and stop bits.

SECTION 2.4: UNDERSTANDING AND USING A LOGIC DIAGRAM FOR SERVICING PURPOSES

The key to understanding logic diagrams begins with understanding truth tables. Figure 2-8 shows the logic symbols and pin configuration for the quad-2-input positive AND gate (7408). If you refer to a manufacturer's data sheet, you'll find the electrical characteristics for this IC are listed as shown (this is only a partial list):

Parameter

$V_{in}(1)$ Logical 1 input voltage required at both input terminals to ensure logical 1 level at output.

$V_{in}(0)$ Logical 0 input voltage required at either input terminal to ensure logical 0 level at output.

$V_{out}(1)$ Logical 1 output voltage.

$V_{out}(0)$ Logical 0 output voltage.

Now, let's see if we can express the electrical characteristics of the 7408 IC in tabular form. Because all four AND gates work exactly the same, we only need to examine one to find out how each will perform. For instance, the AND gate in the lower left-hand corner, with input pins 1 and 2 and output pin 3, will do. Now, if we place the electrical characteristics, given by the manufacturer, in tabular form, the end result is as shown in Table 2-5.

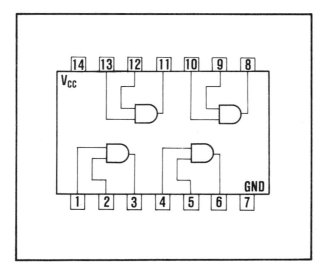

Figure 2-8: Pin configuration and logic symbols for the four AND gates contained in a 7408 IC.

INPUT		OUTPUT
PIN 1	PIN 2	PIN 3
1	1	1
0	1	0
1	0	0
0	0	0

Table 2-5: Table showing all the possible input/output logic levels of an AND gate in a 7408 IC. This is called a *truth table*.

Looking at the truth table for the AND gate shown in Table 2-5, we can derive a Boolean algebra formula for the device. Notice, it takes a logic level 1 on both inputs to produce a logic level 1 on the output. In other words, 1 and 1 = 1 or, A · B = C, in Boolean algebra.

Practical Guide to Servicing OR Gates

Next, the pin configuration and logic symbols for the 7432 quadruple 2-input positive OR gates are shown in Figure 2-9. The 7432 provides four 2-input or logic functions. Each gate may be used individually or connected serially to provide an equivalent 5-input OR function.

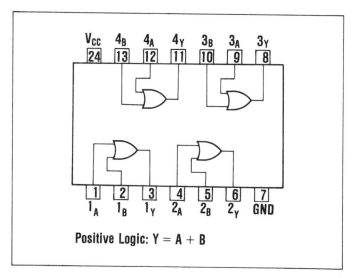

Figure 2-9: Pin configuration and logic symbols for a quadruple 2-input positive OR gate IC (7432).

The OR gate is somewhat different from an AND gate. For example, if you inject an input using a logic level 1 to one of the gates shown in Figure 2-9, the output will be a logic level 1. The formula for the OR gate states that input pins 1 or 2, 4 or 5, 13 or 12, 10 or 9, equal an output. In Boolean algebra: A + B = C. The symbol (+) stands for OR in Boolean algebra. The truth table for a single OR gate (positive logic) is shown in Table 2-6.

Guidelines for Servicing NAND Gates

Because a NAND gate is one of the most often used gates in digital circuitry, it's a good idea to understand how it works from all angles. First, let's look at the quad 2-input NAND gates in a 7400 IC. These are shown in Figure 2-10.

INPUT		OUTPUT
PIN 1	PIN 2	PIN 3
0	0	0
0	1	1
1	0	1
1	1	1

Table 2-6: Truth table for an individual OR gate in the IC shown in Figure 2-9.

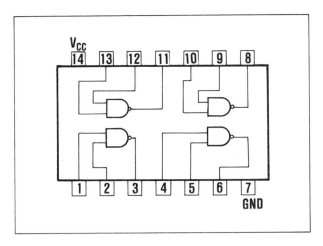

Figure 2-10: Quad 2-input NAND gate IC (7400).

Referring to a single NAND gate logic symbol in Figure 2-10, you'll notice that a NAND gate is actually an AND gate followed by an inverter. The truth table for any one of the NAND gates in Figure 2-10 is shown in Table 2-7.

The operation of a NAND gate is represented by the equation $A = \overline{B \cdot C}$ and is read as output A is the result of $\overline{B}$ and $\overline{C}$ operating at the input of the NAND gate but inverted at the output. The solid bar means inversion.

How to Analyze Exclusive OR Gate Logic Circuits

The OR gates shown in Figure 2-9 are called *inclusive* OR gates because,

INPUT		OUTPUT
PIN 1	PIN 2	PIN 3
0	0	1
0	1	1
1	0	1
1	1	0

Table 2-7: A single 2-input NAND gate truth table using the first NAND gate pin numbers 1, 2, and 3 in Figure 2-10, as an example.

as the truth table (Table 2-6) shows, any input (or all inputs) that contains a logic level 1 will produce a logic level 1 at the output. Now, what do you do when you need an OR gate to produce a logic level 1 when, and only when, one input is a logic level 1? To answer this question, let's first draw a truth table of what we want. Table 2-8 shows the desired operational conditions of the IC we would like.

INPUTS		OUTPUT
AT A	AT B	AT C
0	0	0
0	1	1
1	0	1
1	1	0

Table 2-8: Truth table of an IC requirement explained in the text.

As you have probably already guessed, it happens that an *exclusive* OR gate has exactly the same truth tables as that shown in Table 2-8. Now, all we need is the pin configuration and identifying number of an exclusive OR gate IC. You will notice when referring to the truth table that you must place a 1 on either input A or B (but not both) in order to have a 1 on the output. To say essentially the same thing using a formula, $A\overline{B} + \overline{A}B = C$. Figure 2-11 shows a quad 2-input exclusive OR gate.

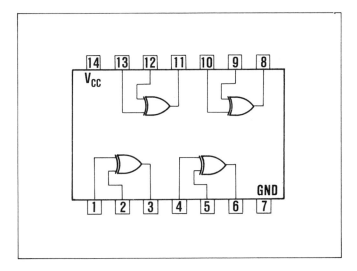

Figure 2-11: Quad 2-input exclusive OR gate.

Testing the Operation of a NOR gate

Now, let's see what happens when an OR logic gate is followed by an inverter. Figure 2-12 shows a low-cost quad 2-input positive NOR gate IC (7402) that is a combination of OR gates followed by inverters.

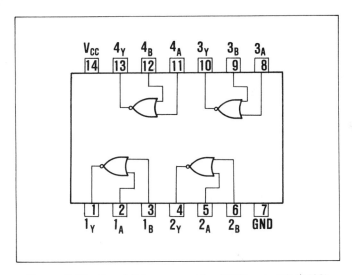

Figure 2-12: Quad 2-input positive NOR gate IC (7402).

When servicing this IC, you'll find that when you place both inputs (any one of the single gates) at logic level 0, the output should be a logic level 1. If either of the inputs is logic level 1, you should expect to find the output at logic level 0. The operation of one NOR gate is represented in the truth table shown in Table 2-9.

INPUT		OUTPUT
A	B	$Y = \overline{A + B}$
0	0	1
0	1	0
1	0	0
1	1	0

Table 2-9: Truth table for the operation of a single positive NOR gate. See Figure 2-12 for the logic symbols and IC package of a 7402 NOR gate.

SECTION 2.5: TESTING LOGIC ICs (GATES)

Perhaps the simplest (and least expensive) way to test digital circuits is to use a digital pulser and a logic monitor. *Note:* You will also need a dc power supply with an output of 15 Vdc to perform the example tests we will cover. To begin, we'll say a few words about the pulser and monitor. However, you should also read Section 2.14 before making any of the following tests, especially if you are inexperienced.

How to Use a Digital Pulser

A digital pulser is frequently called a *logic pulser* but whatever it is called, it's simply a pulse generator that usually provides either a single-pulse and/or a continuous-pulse stream. In passing, we should mention that, although a pulser is not absolutely essential to some troubleshooting jobs, it is a valuable troubleshooting tool. In general, most conventional pulsers provide somewhere near 700 mA of current, which is all you'll need to force most ICs to change states. Some commercial models are pencil-size and will pulse any family of digital circuits. To use a digital pulser, first connect the ground clip to ground and then place the tip of the pulser to the input of the suspect digital IC, etc. Figure 2-13 gives a general idea of the pulser placement.

You will generally find that a pulser has a trigger button. When you press it, you should see the logic element you are testing change state (some digital logic testers emit an audible tone; high tone for the high logic and low tone for the low logic state).

You can signal-trace with a logic pulser and logic probe by simply

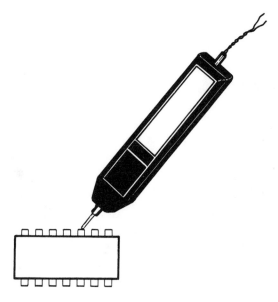

Figure 2-13: Digital pulser being used to pulse a digital element.

following the pulser's signal with the probe. For example, if you find current flowing from one gate to another, but there are no pulses, then either an input or an output is shorted. The symptom you'll first notice in a situation like this is that the pulser will have no effect when you touch the tip to the shorted circuit because, when there is a short to ground, even a pulser cannot change the state of the circuit under test. *Note:* Remember, a signal line may be at logic 1 or logic 0 and be correct, depending on whether you are checking positive or negative. In other words, a certain circuit may be shorted to a logic high.

As an example of how you can use a pulser and logic probe, let's say you want to check a single NAND gate. When you are testing each gate, one of the inputs must be tied to a proper input level (in this example, V_{cc}, which is a high). To test this gate using a pulser and logic probe, your connections would be as shown in Figure 2-14.

The pulser is placed at pin 2, while the probe is placed at the output, pin 3. You should read a logic level 1 on the probe, since the pulser is normally low until you press the trigger key. Pressing the logic pulser trigger switch should cause the probe to read a change of state.

Observing Signals with a Logic Monitor

A logic monitor (also called a *logic clip*) automatically displays static and dynamic logic states of digital ICs. Most of them work with DTL, HTL, TTL, and CMOS dual-in-line package (DIP) ICs. Typically, they have 16 LED displays. One such instrument is shown in Figure 2-15.

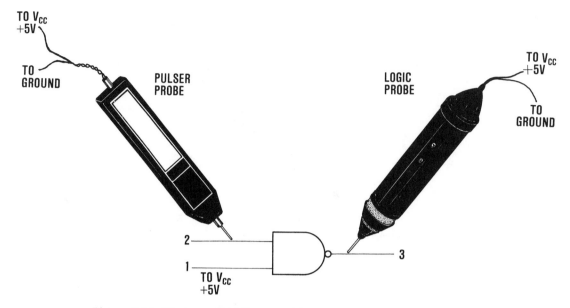

Figure 2-14: Testing a NAND gate with a logic pulser and logic probe.

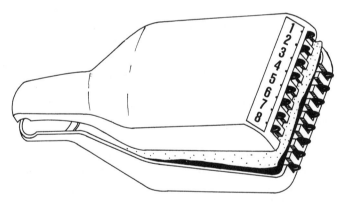

Figure 2-15: Logic monitor that can check all pins of a DIP digital IC.

The monitor is simply piggybacked onto either a 14-or-16-pin DIP IC. It draws power from the circuit under test. If you have made good contact between all monitor clips and IC pins, the monitor LEDs will indicate the logic levels (when an LED is ON, it indicates logic level 1 and OFF is logic 0). Basically, the monitor is nothing but 14 or 16 logic probes all connected and functioning together. However, there are no special external power connections, as with a logic probe. The logic monitor, like the logic probe, is best suited for low-frequency operation. Therefore, about the same problems are encountered when using the monitor as when using a probe.

Besides the pulser, monitor, and dc power supply, you'll also need the spec sheet for the IC you are testing. To give you an idea of how to test a certain type of IC, we will use a MC14011D produced by Motorola.

Testing the Unit as a Control Gate

Comments:

There are four individual NAND gates in this IC. Use the same procedure to check the other three. *Important:* Always connect all *unused* pins to either pin 7 or pin 14 during each gate test.

Procedure:

Step 1. Connect all pins *except* pins 1, 2, and 3 (the first gate in Figure 2-16) to either pin 7 or 14.

Step 2. Connect V_{DD} and V_{SS} (+3 to 15 Vdc) to pins 14 (+V_{DD}) and pin 7 (V_{SS}). A 9-volt battery serves as a power source, if desired.

(A)

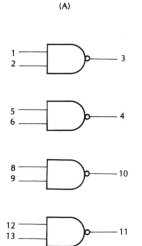

(B)

MAXIMUM RATINGS (Voltages referenced to V_{SS})

RATING	SYMBOL	VALUE	UNIT
DC Supply Voltage	V_{DD}	− 0.5 to 18	V_{dc}
Input Voltage (All inputs)	V_{in}	− 0.5 to V_{DD} + 0.5	V_{dc}
DC Current Drain Per Pin	I	10	mAdc
Operating Temp. Range- AL Device CL/CP Device	T_A	− 55 to + 125 − 40 to + 85	°C
Storage Temp. Range	T_{stg}	− 65 to + 150	°C

Unused inputs must always be tied to an appropriate logic voltage level (e.g., either V_{SS} or V_{DD})

Figure 2-16: (A) Logic diagram for the quad 2-input NAND gate MC14011B. (B) Maximum ratings for this IC.

INPUTS		OUTPUTS
Pins		Pin
1	2	3
L	L	H
L	H	H
H	L	H
H	H	L

Table 2-10: Truth table for testing a MC14011B single control gate.

Step 3. Use truth table 2-10, using the listed inputs and resulting outputs, to check the gate (L = low, H = high).

Testing the MC14011B Wired as an Inverter

Comments:

By using the wiring diagram shown in Figure 2-17 and connecting pins 1 and 2, 5 and 6, 8 and 9, 12 and 13, it is possible to construct four separate inverters with this single IC.

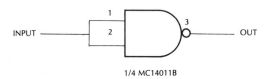

1/4 MC14011B

Figure 2-17: Wiring diagram for an inverter.

Procedure:

Step 1. Connect all pins *except* pins 1, 2, and 3 to either pin 7 or pin 14.

Step 2. Connect V_{DD} and V_{SS} to power source.

Step 3. Use truth table 2-11 to check the inverter gate (L = low, H = high).

INPUTS		OUTPUTS
Pins		Pin
1	2	3
L	L	H
H	H	L

Table 2-11: Truth table for testing the inverter shown in Figure 2-17.

Testing the IC Wired as an AND Gate

Comments:

A formula for an AND gate states that "A AND B = C" (pins 1 AND 2 = A AND B in Table 2-12). The symbol for multiplication (·) stands for AND in Boolean algebra. The table shows the resulting output condition for all possible inputs and the formula is $A \cdot B = C$, or (using pin numbers) $1 \cdot 2 = 4$.

Procedure:

Step 1. Connect all pins *except* 1, 2, 3, 4, 5, and 6 to either pin 7 or 14.

Step 2. Connect V_{DD} and V_{SS} to the power source and see maximum ratings for this IC (given in Figure 2-16).

Step 3. Use truth table 2-12 to check the AND gate (L = high, H = low).

INPUTS		OUTPUTS
Pins		Pin
1 (A)	2 (B)	3 (C)
L	L	L
L	H	L
H	L	L
H	H	H

Table 2-12: Truth table for testing the AND gate shown in Figure 2-18.

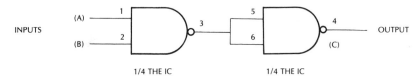

Figure 2:18: Wiring diagram for constructing an AND gate using a MC14011B.

The formula for the previous test (constructing an inverter) is A + B = C. To put it another way, it takes two lows to produce a high. On the other hand, it takes two highs to produce a low. By referring to Table 2-11, you can see that this is true.

Testing the Unit as an OR Gate

Comments:

You should find that any high input (input A or B) to an OR gate will produce a high on the output. The symbol for addition (+) stands for OR in Boolean Algebra.

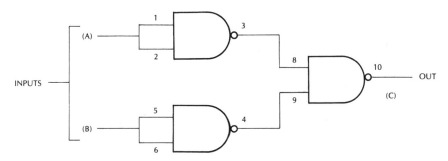

Figure 2-19: Wiring diagram for constructing an OR gate using a MC14011B.

Procedure:

Step 1. Connect pins 11, 12, and 13 to pin 14 (V_{DD}). See Figure 2-19.

Step 2. Connect V_{DD} and V_{SS} to the power source and see maximum ratings for this IC (given in Figure 2-16).

Step 3. Use truth table 2-13 to check the OR gate (L = low, H = high).

INPUTS		OUTPUT
Pins		Pin
1, 2 (A)	5, 6 (B)	10 (C)
L	L	L
L	H	H
H	L	H
H	H	H

Table 2-13: Truth table for an OR gate. The formula for an OR gate states that "A OR B = C" (A + B = C).

COMBINATION GATES

Testing the Combination Gate

Comments:

All types of logic gates, NOR (MC14001B), OR (MC14071B), and AND (MC14081B), can be combined together to form switching arrangements to perform certain operations or functions. We have just described the AND-to-OR gate network. However, there are several other combinations using different multiple gate ICs that you may encounter while servicing digital equipment.

Procedure:

Step 1. Connect all unused pins to either pin 14 (V_{DD}) or pin 7 (V_{SS}). See Figure 2-20.

Step 2. Connect V_{DD} and V_{SS} to the power source and see maximum ratings for this IC (given in Figure 2-16).

INPUTS		OUTPUTS
Pins		Pin
1,2	5, 6	10
X, X	H, H	H
H, H	X, X	H
H, H,	H, H	H

Table 2-14: Truth table for combination (AND-OR). The formula for this combination gate is output at pin 3 = A · B, output at pin 4 = CD, output at pin 10 = AB + CD.

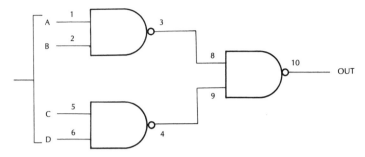

Figure 2-20: Wiring diagram for a combination gate (AND-OR) using a MC14011B.

Step 3. Use truth table 2-14 to check the combination gate (L = low, H = high, X = don't care).

THE MC14011B AS A NOR GATE

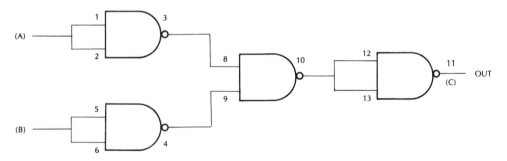

Figure 2-21: Wiring diagram for a NOR gate using a MC14011B.

Testing the NOR Gate

Comments:

The NOR gate is actually an OR gate (see testing the OR gate) followed by an inverter, and may be read as "NOT-OR," hence the term "NOR." By referring to Table 2-15, you will note that when both inputs are low, the output of the NOR gate will be high. The formula reads "A Not and B Not = C" (A + B = C). Incidentally, when the inputs of the NOR gate are connected together, it becomes an inverter.

Procedure:

Step 1. Connect power to V_{DD} and V_{SS}. Do not exceed the maximum ratings given in Figure 2-16.

INPUTS		OUTPUTS
Pins		Pin
1, 2 (A)	5, 6 (B)	10 (C)
L	L	H
L	H	L
H	L	L
H	L	L

Table 2:15: Truth table for a NOR gate.

Step 2. Use truth table 2-15 to check the inputs and outputs (L = low, H = high).

AN EXCLUSIVE-OR GATE USING A MC14011B

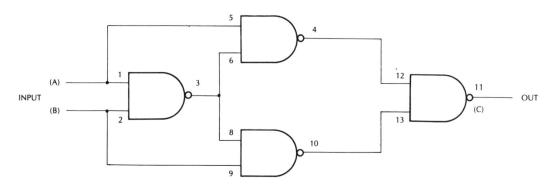

Figure 2-22: Wiring diagram for the Exclusive-OR gate.

INPUTS		OUTPUTS
Pins		Pin
1, 5 (A)	2, 9 (B)	11 (C)
L	L	L
L	H	H
H	L	H
H	H	L

Table 2-16: Truth table for the Exclusive-OR gate. Formula: AB + AB = C.

Testing the Exclusive-OR Gate:

Comments:

Some digital circuits require an OR gate to produce a high on the output when only one input is high. This type of OR gate is called an *Exclusive OR gate*. The formulas given (Table 2-16), read "A and B NOT or A Not and B = C." The truth table shows that only one input with a high will produce an out of a high.

Procedure:

Step 1. Connect power to pins 14 (V_{DD}) and 7 (V_{SS}).

Step 2. Use truth table 2-16 to check inputs and output (L = low, H = high).

A 4-INPUT NAND GATE USING TWO MC14011Bs

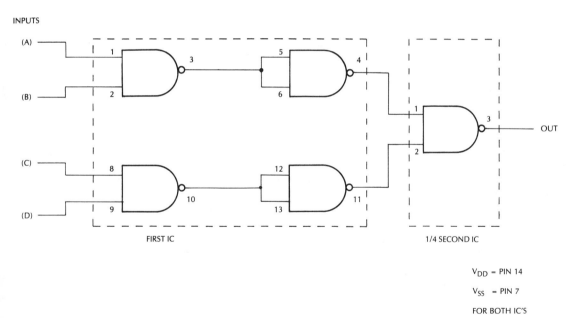

Figure 2-23: Using two MC14011Bs to construct a 4-input NAND gate.

Testing the 4-Input NAND Gate

Comments:

This test uses a quad 2-input NAND gate to construct a single 4-input NAND gate. However, there are 4-input NAND gates in a single package; for example, the dual 4-input NAND gate, MC140128. Of course, this IC uses exactly the same truth table for both of its gates.

Other quad 2-input gates that you may encounter are the NOR gate MC14001B, the OR gate MC14071B, and the AND gate MC14081B. In fact, there is an entire line of this type of gate called "CMOS B-series gates," with 2-inputs, 3-inputs, 4-inputs, and 8-inputs.

Procedure:

Step 1. Connect all unused pins to either pin 14 (V_{DD}) or pin 7 (V_{SS}).

Step 2. Connect the power source to V_{DD} and V_{SS}.

Step 3. Use truth table 2-17 to check the inputs and output (L = low, H = high, X = don't care).

INPUTS				OUTPUT
Pins				Pin
1	2	8	9	3
L	X	X	X	H
X	L	X	X	H
X	X	L	X	H
X	X	X	L	H
H	H	H	H	L

Table 2-17: Truth table for a 4-input NAND gate.

Decoder ICs

There are several types of decoders. For example, Motorola's SN74LS145 IC, a 1-of-10 decoder/driver (open collector) is designed to accept BCD inputs and provide outputs to drive 10-digit incandescent displays. This is a low power *Schottky* TTL device.

What this last statement means is that this IC's internal design does not use the multi-emitter input structure that originally gave TTL its name. Schottky diodes perform the AND function we are about to describe, using TTL decoders. However, in general, all decoders are constructed using an array of inverter and gates, as you will see.

WHAT YOU SHOULD KNOW ABOUT BINARY-TO-DECIMAL DECODER ICs

The basic circuit used to make a decoder is called a *matrix*. For our purpose, a matrix is an array of logic gate connections constructed inside an IC connecting integrated inverters and AND or NAND gates. The logic diagram of a BCD-to-decimal decoder (7442) is shown, with its truth table, in Figure 2-24.

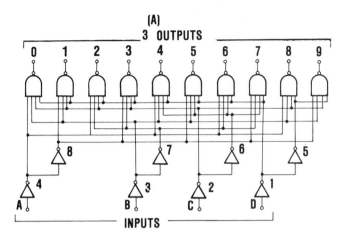

Figure 2-24: 7442 (A) logic diagram (B) truth table.

BCD INPUT				DECIMAL OUTPUT									
D	C	B	A	0	1	2	3	4	5	6	7	8	9
0	0	0	0	0	1	1	1	1	1	1	1	1	1
0	0	0	1	1	0	1	1	1	1	1	1	1	1
0	0	1	0	1	1	0	1	1	1	1	1	1	1
0	0	1	1	1	1	1	0	1	1	1	1	1	1
0	1	0	0	1	1	1	1	0	1	1	1	1	1
0	1	0	1	1	1	1	1	1	0	1	1	1	1
0	1	1	0	1	1	1	1	1	1	0	1	1	1
0	1	1	1	1	1	1	1	1	1	1	0	1	1
1	0	0	0	1	1	1	1	1	1	1	1	0	1
1	0	0	1	1	1	1	1	1	1	1	1	1	0
1	0	1	0	1	1	1	1	1	1	1	1	1	1
1	0	1	1	1	1	1	1	1	1	1	1	1	1
1	1	0	0	1	1	1	1	1	1	1	1	1	1
1	1	0	1	1	1	1	1	1	1	1	1	1	1
1	1	1	0	1	1	1	1	1	1	1	1	1	1
1	1	1	1	1	1	1	1	1	1	1	1	1	1

Referring to the input table shown in Figure 2-24, you will notice that the inputs are labeled D, C, B, and A, rather than A, B, C, and D, as in some other IC truth tables. This notation is used for most IC decoder devices. The D is used to indicate the most significant bit (MSB) of the input and the A, the least significant bit (LSB). To give you an idea of what is taking place in a 7442 decoder, let's chase an input signal (binary word) through the device.

Starting off with DCBA = 0000, you'll see that all outputs from the inverters (1, 2, 3, and 4) are complemented, i.e., $\overline{D}$, $\overline{C}$, $\overline{B}$, and $\overline{A}$. Therefore, the outputs are all at logic 1. Next, the second set of inverters (5, 6, 7, and 8) receive the inverted binary word and invert it again, resulting in outputs of D, C, B and A, which are fed to the matrix as logic 0s. What this all adds up to is that there are two words (1111 and 0000) being fed to the inputs of the NAND gates. Now, if you trace the connections of the NAND gate on the extreme left (output number 0), you will find that all the *inputs* to this gate are logic 1. Therefore, its output would switch to logic 0. Under these same conditions, you'll find that at least one other input on all other NAND gates is logic 0. For this reason, the truth table shows all other decoder outputs are at a logic 1 level.

The 7442 decodes a 4-bit BCD number to one of ten outputs. In the example we just followed through, it was shown that the decoder indicated that the input was the BCD code for decimal number 0 when the 0 output switch was set to logic 0. This same information is given in the truth table. Also, notice that the truth table shows that each input condition causes just

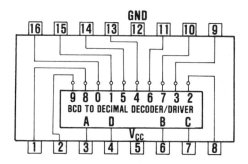

INPUT				OUTPUT ON*
D	C	B	A	
0	0	0	0	0
0	0	0	1	1
0	0	1	0	2
0	0	1	1	3
0	1	0	0	4
0	1	0	1	5
0	1	1	0	6
0	1	1	1	7
1	0	0	0	8
1	0	0	1	9

*ALL OTHER INPUTS ARE OFF

Figure 2-25: BCD-to-decimal decoder/driver 7441 pin configurations and truth table.

one output to switch to logic 0. This is the output corresponding to the decimal equivalent number of the BCD code.

Question: What if you wanted to make up a decoder that sinks the decimal output to logic 1 instead of sourcing to logic 0? You can do this very easily. Simply replace the NAND gates shown in Figure 2-24 with AND gates. Your new truth table would now show logic 0s in place of each logic 1 shown in the truth table in Figure 2-24. All logic 0s would be replaced with logic 1s and that's all there is to it.

Another question: What if you wanted to drive a high current load with a BCD-to-decimal decoder? In this case, a 7445 or 74145 might be your best bet. These ICs have the same pin layouts as the 7442, but their outputs feature high current capabilities (80 mA). The minimum output breakdown voltage for the 7445 is 30 volts and , for the 74145 it's 15 volts. Incidentally, the 7442 outputs are buffered to sink 16 mA.

Two other BCD-to-decimal decoders that have been designed to provide the necessary high-voltage characteristics required for driving gas-filled cold-cathode indicator tubes, relays, or other high-voltage interface circuitry are the 7441 and 74141. Figure 2-25 shows the pin configuration and a truth table for the 7441. The voltage on any output is recommended not to exceed a maximum of 70 volts. Recommended supply voltage is 4.75V to 5.25V.

The 7441 and 74141 have the same pin configuration, as is shown in Figure 2-25. However, the 74141 is a BCD-to-decimal decoder/driver with

INPUT				OUTPUT ON*
D	C	B	A	
L	L	L	L	0
L	L	L	H	1
L	L	H	L	2
L	H	H	H	3
L	H	L	L	4
L	H	L	H	5
L	H	H	L	6
L	L	H	H	7
H	L	L	L	8
H	L	L	H	9
H	L	H	L	NONE
H	L	H	H	NONE
H	H	L	L	NONE
H	H	L	H	NONE
H	H	H	L	NONE
H	H	H	H	NONE

H = HIGH LEVEL, L = LOW LEVEL
***ALL OTHER OUTPUTS ARE OFF**

Table 2-18: Truth table for a 74141 BCD-to-decimal decoder/driver with blanking.

blanking, and its truth table is quite a bit different. Table 2-18 is the truth table for the 74141. Full decoding is provided for all possible inputs stated. For binary inputs 10 through 15, all outputs are off. This IC was designed specifically to drive cold-cathode indicator tubes such as a NIXIE (Trademark of Burrough Corp). When the IC is connected to external digital circuits, the automatic blanking control ensures that the NIXIE tubes the IC is driving are not conducting for the time the next BCD coded word is shifted into the decoder. V_{CC} + 5V nom., power dissipation = 55 milliwatts (typically), output voltage is 65 (max.), and supply current V_{CC} (max) = 16 mA.

Binary-to-Octal Decoders

Before we examine the binary-to-octal decoder IC, let's review a bit. To convert an octal number to its binary equivalent, you will remember that in a previous section it was stated that you must convert each digit of the octal number to the binary equivalent by *using three binary digits per octal digit*. Another point: for three octal digits, it is necessary to use nine binary digits. On the other hand, to convert a binary number to its octal equivalent, you must divide the digits of the binary number into groups of three. Starting from the right, add 0 to the left to complete the digits, if necessary. For example:

added zero	010	100	111 (binary)
	2	4	6 (octal)

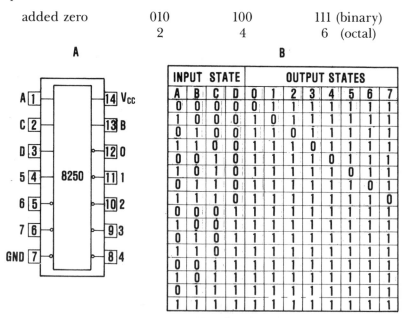

INPUT STATE				OUTPUT STATES							
A	B	C	D	0	1	2	3	4	5	6	7
0	0	0	0	0	1	1	1	1	1	1	1
1	0	0	0	1	0	1	1	1	1	1	1
0	1	0	0	1	1	0	1	1	1	1	1
1	1	0	0	1	1	1	0	1	1	1	1
0	0	1	0	1	1	1	1	0	1	1	1
1	0	1	0	1	1	1	1	1	0	1	1
0	1	1	0	1	1	1	1	1	1	0	1
1	1	1	0	1	1	1	1	1	1	1	0
0	0	0	1	1	1	1	1	1	1	1	1
1	0	0	1	1	1	1	1	1	1	1	1
0	1	0	1	1	1	1	1	1	1	1	1
1	1	0	1	1	1	1	1	1	1	1	1
0	0	1	1	1	1	1	1	1	1	1	1
1	0	1	1	1	1	1	1	1	1	1	1
0	1	1	1	1	1	1	1	1	1	1	1
1	1	1	1	1	1	1	1	1	1	1	1

Figure 2-26: Binary-to-octal decoder (8250). (A) is the pin configuration for a 14-lead dual in-line-molded. (B) is a truth table for the 8250 or MC7250P. The selected output is a logic 0.

Now, from this brief review, it appears that a binary-to-octal decoder would require three lines of input. Figure 2-26 shows that this is true. Our example binary-to-octal decoder, the 8250 or MC7250P, converts three lines of input (A,B, and C) to one-of-eight output. The fourth input (D) is utilized as an inhibit to allow use in larger decoding networks.

This IC accepts a BCD number at inputs A, B, C, D, and provides a low at one of eight outputs, while the other seven outputs remain high. For example, a 1010 input results in output 5 (pin 4) going low. Binary numbers from 1110 to 1111 at the input send all outputs high. See the logic diagram for the 8250 shown in Figure 2-27. Notice that the logic diagram for this IC is very similar to that of the 7442 BCD-to-decimal decoder (and all the rest of the decoders we have discussed). Almost all of these TTL decoders/drivers are much the same except for the internal wiring network, i.e., the matrix.

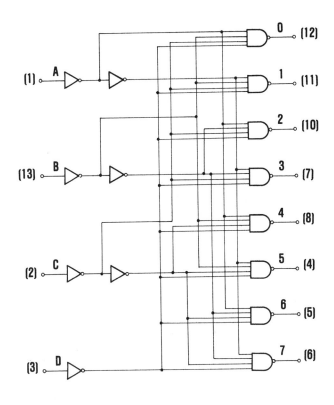

V_{CC} = (14), GND = 7

( ) DENOTES PIN NUMBERS FOR 14 PIN DUAL-IN-LINE PACKAGE

Figure 2-27: Logic diagram for the 8250 binary-to-octal decoder.

SECTION 2.6: WORKING WITH DECODERS/ENCODERS

In the world of the serviceman, encoders are usually thought of as devices that convert binary numbering systems back into other numbering systems. A CMOS/BCD-to-decimal decoder/binary-to-octal decoder IC (the MC14028B) block diagram and truth table are shown in Figure 2-28. Ordinarily, a decoder IC would be used at the output terminals of a digital system, but an encoder usually is used at the inputs.

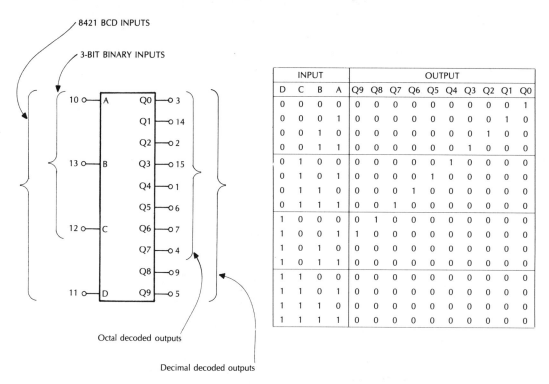

INPUT				OUTPUT									
D	C	B	A	Q9	Q8	Q7	Q6	Q5	Q4	Q3	Q2	Q1	Q0
0	0	0	0	0	0	0	0	0	0	0	0	0	1
0	0	0	1	0	0	0	0	0	0	0	0	1	0
0	0	1	0	0	0	0	0	0	0	0	1	0	0
0	0	1	1	0	0	0	0	0	0	1	0	0	0
0	1	0	0	0	0	0	0	0	1	0	0	0	0
0	1	0	1	0	0	0	0	1	0	0	0	0	0
0	1	1	0	0	0	0	1	0	0	0	0	0	0
0	1	1	1	0	0	1	0	0	0	0	0	0	0
1	0	0	0	0	1	0	0	0	0	0	0	0	0
1	0	0	1	1	0	0	0	0	0	0	0	0	1
1	0	1	0	0	0	0	0	0	0	0	0	0	0
1	0	1	1	0	0	0	0	0	0	0	0	0	0
1	1	0	0	0	0	0	0	0	0	0	0	0	0
1	1	0	1	0	0	0	0	0	0	0	0	0	0
1	1	1	0	0	0	0	0	0	0	0	0	0	0
1	1	1	1	0	0	0	0	0	0	0	0	0	0

Figure 2-28: Block diagram and truth table for the BCD-to-decimal decoder/binary-to-octal decoder (MC14028B).

The MC14028B is constructed so that an 8-4-2 BCD code on the four inputs (A, B, C, and D) provides a decimal (one-of-ten) decoded output, while a 3-bit binary input—A, B, and C code output with input D (pin 11) forced to a logic low level. This IC is useful for code conversion, address decoding in computer applications, or such applications as memory selection.

Frequently, you will find the ICs designed as encoders are called *priority encoders*. As the name implies, this device was designed to give a priority to certain inputs. When more bits-per-output are required, the encoder may become very complex. Because of this, it is usual to find a ROM being used as an encoder in computer systems. Two typical encoder ICs of the type we are discussing are the TTL 74147 10-line-to-4-line and the CMOS 14532B 8-bit (eight data inputs, five data outputs) priority encoders. Figure 2-29 shows a truth table and pin configuration for the 74147, and Figure 2-30 is a block diagram and truth table for the MC14532B.

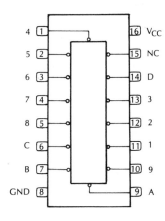

INPUTS									OUTPUTS			
1	2	3	4	5	6	7	8	9	D	C	B	A
H	H	H	H	H	H	H	H	H	H	H	H	H
X	X	X	X	X	X	X	X	L	L	H	H	L
X	X	X	X	X	X	X	L	H	L	H	H	H
X	X	X	X	X	X	L	H	H	H	L	L	L
X	X	X	X	X	L	H	H	H	H	L	L	H
X	X	X	X	L	H	H	H	H	H	L	H	L
X	X	X	L	H	H	H	H	H	H	L	H	H
X	X	L	H	H	H	H	H	H	H	H	L	L
X	L	H	H	H	H	H	H	H	H	H	L	H
L	H	H	H	H	H	H	H	H	H	H	H	L

H = High logic level.   L = Low logic level.
X = Irrelevant.

Figure 2-29: Pin configuration and truth table for a 74147 10-line priority encoder.

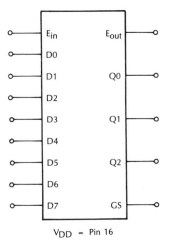

V$_{DD}$ = Pin 16

V$_{SS}$ = Pin 8

INPUT										OUTPUT				
E$_{in}$	D7	D6	D5	D4	D3	D2	D1	D0	GS	Q2	Q1	Q0	E$_{out}$
0	X	X	X	X	X	X	X	X	0	0	0	0	0
1	0	0	0	0	0	0	0	0	0	0	0	0	1
1	1	X	X	X	X	X	X	X	1	1	1	1	0
1	0	1	X	X	X	X	X	X	1	1	1	0	0
1	0	0	1	X	X	X	X	X	1	1	0	1	0
1	0	0	0	1	X	X	X	X	1	1	0	0	0
1	0	0	0	0	1	X	X	X	1	0	1	1	0
1	0	0	0	0	0	1	X	X	1	0	1	0	0
1	0	0	0	0	0	0	1	X	1	0	0	1	0
1	0	0	0	0	0	0	0	1	1	0	0	0	0

X = Don't care.

Figure 2-30: Block diagram and truth table for a MC14532B 8-bit priority encoder.

The 74147 encodes nine data lines to 4-line (8-4-2-1) BCD. The implied decimal zero condition needs no specific input condition because zero is encoded when all nine data lines are at a high logic level. The inputs (E1 and E) are for cascading purposes. They are provided to allow octal expansion without the need of external circuitry. In other words, you can connect one encoder IC to another to expand your operation, if necessary. For this IC, data inputs and outputs are active at the low logic level.

The MC14532B, like other encoders, has a primary function of providing a binary address for the active input with the highest priority. This IC has eight inputs (D0 through D7) and an enable input (E_{in}). You will also notice that five outputs are available; three are address outputs (Q0 through Q2), one group select (GS), and one enable output (E_{out}).

SECTION 2.7: TESTING AN ENCODER IC

As an example of testing an encoder IC, let's assume that you have a 74147 wired up on a breadboard, with switches on the inputs 1 through 9, and LEDs on outputs A through C. Of course V_{cc} (5V) must be applied to pin 16 and ground to pin 8. The actual pin configuration for this IC is shown in Figure 2-31. *Note:* See Section 2.5 for testing instruments.

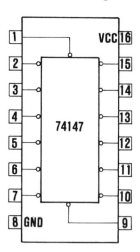

Figure 2-31: Pin configuration for a 74147 that can be used as a guide during breadboard mounting of the IC.

After the IC is properly mounted and power is applied, your next step is to refer to the function table shown in Table 2-19 to see which LED should light when you open or close selected switches on the inputs. First, the

INPUTS									OUTPUTS			
1	2	3	4	5	6	7	8	9	D	C	B	A
H	H	H	H	H	H	H	H	H	H	H	H	H
X	X	X	X	X	X	X	X	L	L	H	H	L
X	X	X	X	X	X	X	L	H	L	H	H	H
X	X	X	X	X	X	L	H	H	H	L	L	L
X	X	X	X	X	L	H	H	H	H	L	L	H
X	X	X	X	L	H	H	H	H	H	L	H	L
X	X	X	L	H	H	H	H	H	H	L	H	H
X	X	L	H		H	H	H	H	H	H	L	L
X	L	H	H	H	H	H	H	H	H	H	L	H
L	H	H	H	H	H	H	H	H	H	H	H	L

H = HIGH LOGIC LEVEL, L = LOW LOGIC LEVEL
X = IRRELEVANT

Table 2-19: Function table for a 74147.

implied decimal zero condition requires no input condition, as zero is encoded when all nine data switches are at a high logic level (see the top line—all Hs—of the function table). All data inputs and outputs are active at the low logic level.

Next, notice that it is irrelevant what input data is entered on inputs 1 through 8 if 9 is set to a logic low level. The output will be 1001 (LHHL) BCD. In our example, two LEDs will be on and two off, i.e., decimal number 9 being 1001. Checking the other lines in the function table, you'll see that the encoder will perform priority decoding of the inputs to ensure that only the highest order data line is encoded. The next line, output LHHH, translates to 1000 binary, or 8 decimal, when input 8 is low and input 9 is high, and each line following can be read in the same manner.

An 8-line-to-3-line decoder, the 74148, pin configuration and function table are shown in Figure 2-32. There are two pins (labeled E1 and E0) on this IC that are not utilized in the same manner as with the 74147. Cascading circuitry (enable input E1 and enable output E0) has been provided to allow octal expansion without the need for external circuitry. As with the 74147, data inputs and outputs are active at the low logic level.

Another priority circuit you may see on your workbench is a MC14530 *majority logic gate*. Before we look at this IC, perhaps we should first explain what a majority gate is. Simply, this gate is a logic element that will produce a logic high on its output if more than half its inputs are at a logic high. On the other hand, the output will be a logic low for other input conditions. *Note:* Actually, the output will follow any given input logic level, high or low. Figure 2-33 shows an example of what we mean. This gate has five inputs and the output is a logic high whenever any three or more of its inputs are at a logic high. It doesn't make any difference which input you make high—any three will cause the output to go high.

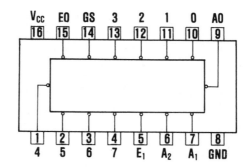

INPUTS									OUTPUTS				
E₁	0	1	2	3	4	5	6	7	A₂	A₁	A₀	GS	EO
H	X	X	X	X	X	X	X	X	H	H	H	H	H
L	H	H	H	H	H	H	H	H	H	H	H	H	L
L	X	X	X	X	X	X	X	L	L	L	L	L	H
L	X	X	X	X	X	X	L	H	L	L	H	L	H
L	X	X	X	X	X	L	H	H	L	H	L	L	H
L	X	X	X	X	L	H	H	H	L	H	H	L	H
L	X	X	X	L	H	H	H	H	H	L	L	L	H
L	X	X	L	H	H	H	H	H	H	L	H	L	H
L	X	L	H	H	H	H	H	H	H	H	L	L	H
L	L	H	H	H	H	H	H	H	H	H	H	L	H

Figure 2-32: Pin configuration and function table for a 74148 8-line-to-3-line priority encoder.

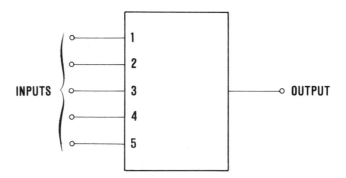

Figure 2-33: Majority logic gate. When the majority of the inputs (1 through 5) assume a given logic level, so does the output.

Now, we'll look at a MC14530. This IC has two majority logic gates (it is a dual 5-input majority logic gate). Figure 2-34 shows a block diagram of one of the gates and its pin connections. The other half is exactly the same except the input pins are 9, 10, 11, 12, and 13. The input pin to its exclusive NOR gate is 14. The output is pin 15.

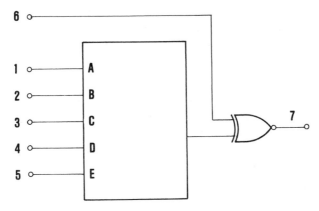

Figure 2-34: One of two 5-input majority logic gates and its NOR gate. The output (pin 7) will be a logic 1 if any three or more inputs are logic 1, and pin 6 is tied to a logic high.

To produce a logic high on both outputs when a majority of inputs are at a logic high, both pins 6 and 14 (the inputs to both exclusive OR gates) should be tied to a high (V_{cc}). Pin 16 is shown in Figure 2-35. But, if you connect pins 6 and 14 to a logic low (ground), the output will be *low* whenever three or more inputs are high. The circuit shown in Figure 2-35 shows how you can

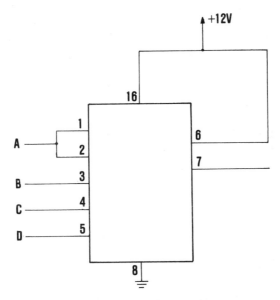

Figure 2-35: Wiring diagram for an MC14530 dual 5-input majority logic gate that will give priority to a selected input.

use the IC to select a certain input over all the rest. To put it another way, give priority to a certain input.

By connecting any two inputs together, you can make the resulting input switch as important as the other inputs (see Figure 2-35). In our example, if A or any one of the other inputs goes high, it will sound an alarm, light an LED, or whatever you desire. Some other tests you might like to try with this IC are:

1. Place a high on pins 1, 2, and 6 (See Figure 2-34). Now, if you place a high on any other pin, you have a 3-input OR gate.
2. Place pin 6 to ground (a logic low). Place pin 1 and 2 at logic high. This will invert the output, as you'll remember. Therefore, you now have a 3-input NOR gate.
3. Place pins 1 and 2 at ground. Place pin 6 at a high. You now have a 3-input AND gate.
4. Again place pins 1 and 2 at ground, also pin 6 to ground. You have a 3-input NAND gate in this case.

You can use this same idea with other ICs; however, *do not short more than one output of any IC during any test.* As a general rule, you can short several *input* pins to ground, *but not output pins!*

SECTION 2.8: READOUTS AND DRIVERS

When we say readouts and drivers, we automatically include the subject of *output interfacing,* particularly if we are breadboarding or redesigning an existing circuit. While all input interface circuits must be designed around the IC's input voltage and current specs, output interface circuits must take into consideration the IC's output voltage and current loading (for example, fan-out rating of a digital IC). *Note:* In case you don't remember, fan-out refers to the number of parallel loads within a certain logic family (TTL, CMOS, MOS, etc.) that can be driven from one output of a digital IC.

One of the most popular output devices for digital circuits is a light-emitting diode (LED). Both the single LED and 7or 8-segment LED numeric display assembly are in wide use. Figure 2-36 shows two discrete LED Indicator circuits you can use; however, 7- or 8-segment displays need some form of decoder to operate properly.

While the technician can use the interface-to-LED circuits shown in Figure 2-36, and they are adequate for breadboarding one-of-a-kind circuits, if he is interested in interfacing several LED displays to a system it's better to use LED driver ICs. For example, the MC75491 (a quad LED segment driver) and the MC75492 (a hex LED digit driver) are designed to interface MOS logic to common cathode LED readouts in serially addressed multidigit displays. ICs such as these two are called *multiple light-emitting (LED) drivers.* The MC75491 and MC75492 are shown in Figure 2-37.

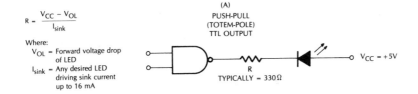

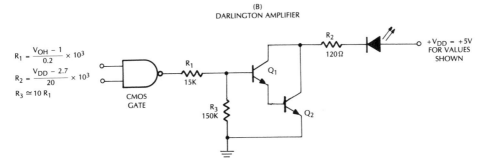

Figure 2-36: (A) TTL-to-LED, (B) CMOS-to-CMOS interface. *Note:* In general, CMOS gates cannot supply enough current to light an LED directly.

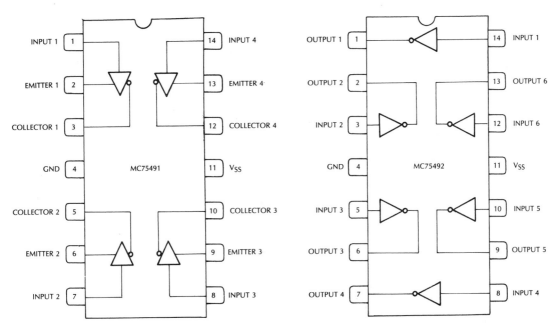

Figure 2-37: Connection diagram for the MC75491 and MC75492 multiple LED drivers.

A typical use of these ICs is for interfacing an MOS calculator IC (chip) to a standard 7-segment multidigit display. Before we look at a demonstration circuit that takes advantage of these multiple LED drivers, let's first review a standard 7-segment layout and associated decimal display that you might encounter while servicing digital equipment.

ICs (such as a calculator chip) used to drive these displays have output pins labeled A, B, C, D, E, F, G, and DP (Decimal Point). In turn, these output pins must turn on certain segments of the 7-segment display to display a certain decimal number. For your convenience, a standard 7-segment decimal display is shown in Figure 2-38.

Now, keep in mind that each digit of a certain decimal display must utilize one entire standard 7-segment display unit for each of the digits displayed. For instance, to display the decimal number 520042 would require six individual 7-segment LED displays to be turned on in accordance with the segment letters shown in Figure 2-38.

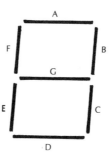

Figure 2-38: Standard 7-segment layout and associated decimal display.

The MOS calculator chip-to-LED interface example circuit suggested by Motorola, and shown in Figure 2-39, is a stripped-down block diagram. This example uses time multiplexing of the individual digits in a visible display to minimize display circuitry. Multiplexing is the best system to use for displays requiring a large number of 7-segment displays. Incidentally, you can also use BCD-to-7-segment decoder ICs (which may be less expensive) where only three or four decimal numbers are to be displayed. The system shown can

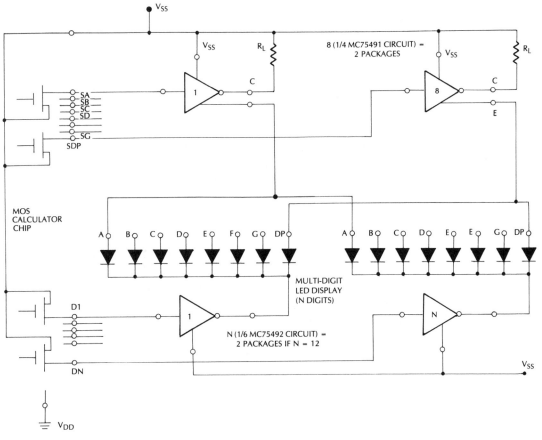

Figure 2-39: MOS calculator chip 7-segment display interface circuit.

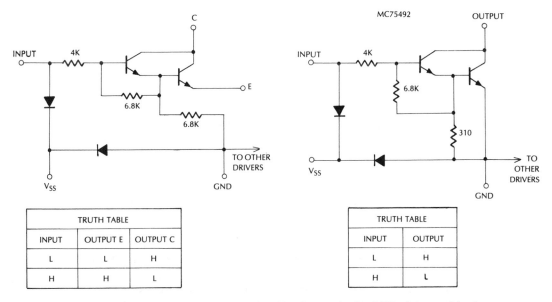

TRUTH TABLE		
INPUT	OUTPUT E	OUTPUT C
L	L	H
H	H	L

TRUTH TABLE	
INPUT	OUTPUT
L	H
H	L

Figure 2-40: Schematic and truth table for a single LED driver with the MC74591/2 IC.

display up to 12 digits with decimal point, using only two MC7459ls and two MC74592 drivers.

Referring to Figures 2-37 and 2-39, notice that the MC75491 ICs have two outputs (C and E, the collector and emitter of an internal transistor), and the MC75492 has only one (the collectors of two internal transistors wired together). The MC75491 driver unit outputs are connected to the individual LED display configuration, therefore it requires logic highs to turn on the segments. An individual driver schematic and truth table for each of these ICs is shown in Figure 2-40.

LED Indicator/CMOS Driver Circuits

In some cases, it is necessary to limit the amount of current through an LED with a limiting resistor (R), as shown in Figure 2-41. Also, you will see that we have included both common cathode and common anode LED circuits in this illustration. The drivers are CMOS ICs with only a single internal unit of the IC being shown.

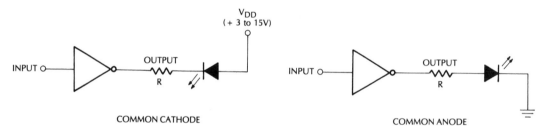

Figure 2-41: LED indicator driver circuits with current-limiting resistors. See Figure 9-10 for other LED driver circuits.

The formula for finding the necessary value of the current-limiting resistor is

$$R = (V_{DD} - V_F)/I_F$$

where V_F is the forward voltage drop measured across the conducting LED (typically, about 1.7 volts). I_F is the desired current flow during the LED operating period (typically 10 mA).

Example:

$$V_{DD} = 11.7 \text{ volts}, V_F = 1.7 \text{ volts}, I_F = 10 \text{ mA}$$
$$R = (11.7 - 1.7)/0.01 = 1000 \text{ ohms}.$$

You will find that 1000 ohms is recommended for many applications that interface an LED to a CMOS IC, and a current-limiting resistor is frequently required.

There are ICs that do not require a current-limiting resistor when their output is driving an LED circuit, as was shown in Figure 2-39. Two methods to connect an LED directly to a CMOS output circuit are shown in Figure 2-42.

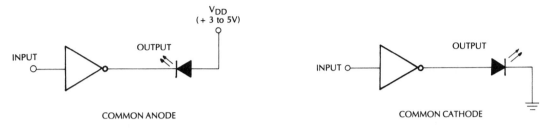

Figure 2-42: How to connect a common anode and common cathode LED circuit to a CMOS IC output terminal with V_{DD} in the 3- to 5-volt range (in this case, no current-limiting resistor is required).

Driving High-Voltage/Current Peripherals

To drive a peripheral often requires more current and/or voltage than an IC or discrete component can deliver. Figure 2-43 shows a circuit that can be used to drive a discrete transistor amplifier circuit that will provide a considerable *current* gain.

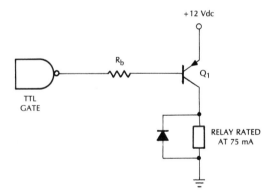

Figure 2-43: Interfacing a low current (16 mA sink current) TTL gate to a relatively high current load. The diode must be included to prevent current surges by the sudden opening and closing of the relay.

As you may know, the sink current level of most TTL gates and many other TTL devices (Schmitt triggers, etc.) is rated at 16 mA. Incidentally, TTL buffers and the like usually drain off much more energy from the system. Therefore, we can take advantage of these fairly large sink currents that these TTL devices drain off at logic 0. In general, you will find that 16 mA is more than enough to drive the transistor you want to use into saturation.

Now, remembering that the exact relationship between collector current and base current (called *beta*) is expressed as

$$h_{FE} = \beta = I_c/I_b,$$

let's assume that the transistor we are using has a beta of at least 12 (a conservative assumption, in almost any case). It follows that our transistor circuit can drive a load up to 192 mA, with a gate sink current of 16 mA (I_b, in the formula).

Since our amplifier has a current gain of 12, the load current (I_L) equals 12 I_{sink}. Knowing this, the value of R_b can be determined if you assume a certain value of load current. The formula, in this case, becomes

$$R_b = V_{CC} - (V_{OL} + V_{eb})/ \ 1/12 \text{ of the load current.}$$

As an example, let's say that we want to drive a relay coil having a current rating of 75 mA and a resistance of 160 ohms. Using our formula and assuming the transistor emitter base voltage drop (forwarding bias voltage drop) and the other circuit voltage drops add up to about 1 volt, and the voltage on the transistor emitter is +12 Vdc as shown, in this case

$$R_b = 12 \times 11/75 \times 10^{-3} = 1.8 \text{ k}\Omega.$$

An IC that may be used to drive relays and lamps, and as a line driver, MOS driver, or as a buffer, is the MC75461, a positive AND gate array. The other ICs are NAND, OR, and NOR gates; i.e., the 62, 63, and 64 respectively, all have exactly the same pin connections as shown for the MC75461. But obviously, the truth tables are not the same. Each of these peripheral drivers contains a pair of TTL gates, with the output of each gate internally connected to the case of a transistor. See Figure 2-44. They are also TTL compatible and have 300 mA (max) output current capability.

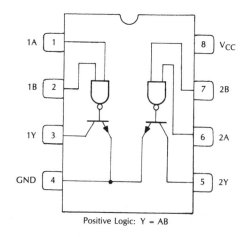

Positive Logic: Y = AB

A	B	Y
L	L	H ("OFF" STATE)
L	H	H ("OFF" STATE)
H	L	H ("OFF" STATE)
H	H	L ("ON" STATE)

Figure 2-44: MC75461 dual high-voltage peripheral drivers.

Testing a Peripheral Driver

You can use the wiring diagram shown in Figure 2-45 to test one of these devices (the MC75461 or similar IC). A few points to remember during this test, or any other, for that matter, are listed here:

1. You must observe all operating restrictions. Reversed polarity, excessive supply voltage, and drawing too much supply current can destroy an IC. For example, Figure 2-45 lists a maximum load current (I_{OL}) of 300 mA, and a maximum load voltage (V_{OL}) of 35Vdc. *Under no circumstance should you exceed these values.*
2. *Never* remove or insert an IC into a socket with power applied to the circuit.
3. In general, do not connect an input pulse generator to an IC with power to the circuit off, *particularly if the IC is a CMOS device.*
4. When working with CMOS ICs, all unused IC inputs should be connected to V_{DD} or V_{SS}. If you are having troubles with a circuit (such as instability or the drawing of too much current), check for unterminated terminals on the IC.

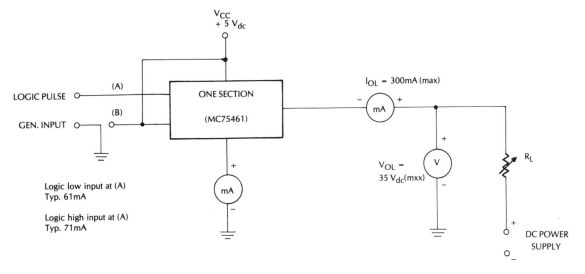

Figure 2-45: Test circuit for a MC75461 IC or similar peripheral driver. Each input is tested separately.

Essentials of Display Decoders

Seven-segment displays (found on many digital readouts, digital instruments, calculators, etc.) usually need a BCD-to-7-segment decoder/driver to operate correctly. The standard TTL family (7446, 7, and 8) BCD-to-7-segment decoder/driver ICs all have the same pin configuration. See Figure 2-46

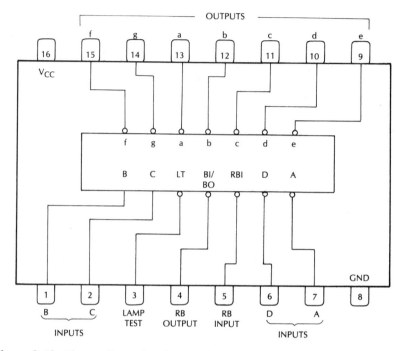

Figure 2-46: Pin configuration for standard TTL BCD-to-7-segment decoder/ driver ICs.

A block diagram for the CMOS BCD-to-7-segment decoder MC14558B is shown in Figure 2-47. This IC decodes 4-bit binary coded decimal data, dependent upon the state of the auxiliary inputs enable (pin 3) and RBI (pin 5), and provides an active high 7-segment output for a display driver. Notice,

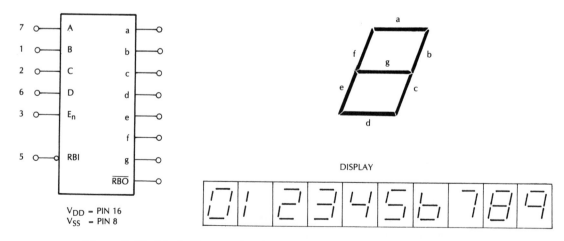

Figure 2-47: Block diagram for the MC14558B BCD-to-7-segment decoder.

the outputs are labeled a, b, c, d, e, f, and g (this is standard for all these decoders, both TTL and CMOS). Each of the lettered outputs corresponds to one segment in the actual display device. See the lettered figure 8 next to the block diagram in the illustration. A truth table is shown in Table 2-20.

INPUTS						OUTPUTS*								DISPLAY
E_{in}	$\overline{RBI}$	D	C	B	A	a	b	c	d	e	f	g	RBO	
Pin 3	Pin 5	Pin 6	Pin 2	Pin 1	Pin 7	Pin 13	Pin 12	Pin 11	Pin 10	Pin 9	Pin 15	Pin 14	Pin 4	
1	1	0	0	0	0	1	1	1	1	1	1	0	1	*0*
1	X	0	0	0	1	0	0	0	0	1	1	0	1	*1*
1	X	0	0	1	0	1	1	0	1	1	0	1	1	*2*
1	X	0	0	1	1	1	1	1	1	0	0	1	1	*3*
1	X	0	1	0	0	0	1	1	0	0	1	1	1	*4*
1	X	0	1	0	1	1	0	1	1	0	1	1	1	*5*
1	X	0	1	1	0	0	0	1	1	1	1	1	1	*6*
1	X	0	1	1	1	1	1	1	0	0	0	0	1	*7*
1	X	1	0	0	0	1	1	1	1	1	1	1	1	*8*
1	X	1	0	0	1	1	1	1	0	0	1	1	1	*9*
1	0	0	0	0	0	0	0	0	0	0	0	0	0	BLANK
0	0	X	X	X	X	1	1	1	1	1	1	1	0	*8*
0	1	X	X	X	X	0	0	0	0	0	0	0	1	BLANK

* All non-valid BCD input codes produce a blank display.   X = Don't care.

Table 2-20: Truth table for the MC14558B BCD-to-7-segment decoder. This example is typical of the inputs required for these decoders, whether TTL or CMOS devices.

Each input/output condition for a particular numerical display is shown, except for the auxiliary inputs RBI = ripple blocking input, and RBO = ripple blocking output. To perform a lamp test (check all seven outputs simultaneously), you would place a 0 on pin 3, 0 on pin 5, X (don't care) on BCD input code, and 0 on pin 4. Also, you can block all segments by placing a high on pin 3 and a low on pin 5, BCD inputs, and pin 4. To perform all numerical displays requires, as explained, the setup shown in Table 2-20.

SECTION 2.9: PULSE CIRCUIT ICs

Digital circuits are entirely useless without a good, clean source of trigger pulses. Multivibrators are oscillators that primarily produce pulse waveforms that are important to digital electronics. There are numerous multivibrator ICs. For example, there is the TTL family, 74121 monostable, and the 74122 retriggerable monostable. Two in the CMOS family are the more recent MC14528B dual retriggerable/resettable monostable and the MC14538B dual precision retriggerable/resettable monostable. Notice that each of our example IC multivibrators is referred to as *monostable*.

1. Astable Multivibrator (Free-Running)

This circuit has two momentarily stable states. It switches rapidly from one state to the other (either a logic high or low). The output signal is a square-edge waveform used as a clock pulse in digital applications—computers, digital games, etc. Figure 2-48 illustrates a simple transistor astable multivibrator.

Note: To make this explanation as simple as possible, we have deleted the resistor-capacitor timing components normally found in the transistor base-to-collector feedback circuit.

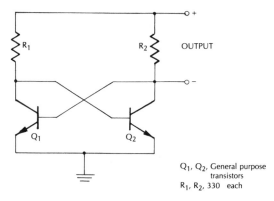

Figure 2-48: A basic bipolar transistor astable multivibrator circuit.

The simple basic circuit shown in Figure 2-48 lacks the precision of the more sophisticated ICs described in the beginning of this section (Motorola's MC14538B), but is certainly much easier to analyze. To begin with, because of slight manufacturing differences, transistors Q1 and Q2 will not conduct the same amount of current when power is first applied to the circuit.

Let's assume that Q1 conducts more current than Q2. Under this condition, the collector voltage of Q1 will initially keep Q2 in a cut-off mode of operation because, as you can see, it is directly wired to the base of Q2. Nevertheless, as Q1 conducts more and more, its collector voltage will drop more and more, and Q2 will be allowed to start conducting. Now that Q2 is turned on, the initial high collector voltage (notice, this collector voltage is also being applied to the base of Q1), will cut Q1 off. But as the current through Q2 increases, the collector voltage decreases. In fact, it will reach a point where Q1 is again conducting and Q2 is turned off.

In summary, the circuit switches back and forth (between Q1 and Q2) between two states (on and off), providing a square waveform output. Incidentally, you can take the output signal from either the collector of Q2, as shown, or the collector of Q1. Now, with this basic understanding of how multivibrators work, let's examine the other types.

2. Monostable Multivibrator IC

Also called *mono, one-shot, single-shot,* or *start-stop.* This is a circuit that has only one stable state. However, it can be triggered to change to another state (logic high or low) for a predetermined period of time. At the end of this time, it will return to its original stable state. You will find that monostable multivibrators are used in applications where delaying or reshaping of pulses is desired. Figure 2-49 shows a logic diagram for one-half of a CMOS dual monostable multivibrator (MC14528B).

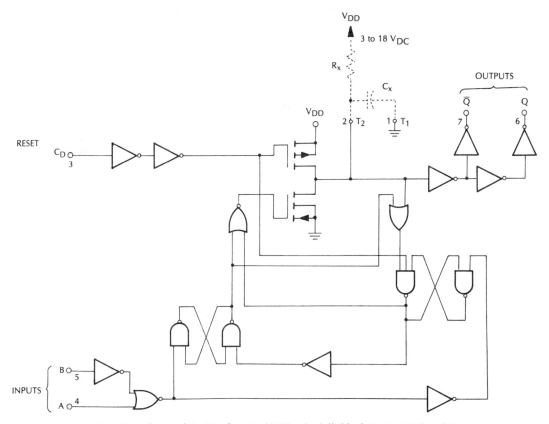

Note: Externally ground pins 1 (as shown) and 15 (the other half of the device) to pin 8. R_x and C_x are external components. V_{DD} = pin 16, V_{SS} = pin 8.

Figure 2-49: Logic diagram for one-half of a CMOS dual monostable multivibrator.

As previously mentioned, a mono can be triggered to change state for a predetermined period of time. The MC14528B can be triggered from either edge of an input pulse. The output will be initiated on the positive-or negative-going edge of the input waveform, depending upon which input is used, A or B (see Figure 2-49).

The time delay of the mono is mainly dependent upon the time constant of the external timing components C_x and R_x, shown as dashed lines in Figure 2-49. Varying these two components will produce an output pulse over a wide range of widths, depending also on which value V_{DD} you used (5, 10, or 15 volts).

You can choose either a negative- or a positive-going output pulse. A positive Q output is available at pin 6, while a complementing $\overline{Q}$ output is at pin 7. When pin 6 is high, inverted output pin 7 is low and vice versa. To place the IC into operation you could apply a positive-going pulse on input A and B (pins 4 and 5) and one-half the device, a negative-going pulse on input C_D (reset), and a negative-going pulse on the other half of the IC (pins 11 and 12). A block diagram of the entire IC is shown in Figure 2-50.

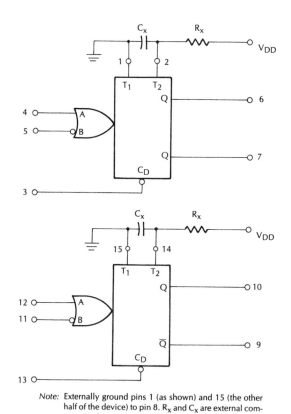

Note: Externally ground pins 1 (as shown) and 15 (the other half of the device) to pin 8. R_x and C_x are external components. V_{DD} = pin 16, V_{SS} = pin 8.

Figure 2-50: A block diagram of the MC14528B dual monostable multivibrator.

Basically, all multivibrators are *square-wave* generators. There are other ICs than the multivibrator ICs we have just used as examples that can be used to create a square-wave generator; for example, a hex Schmitt trigger such as the MC14584.

Testing a Hex Schmitt Trigger as a Square-Wave Generator

Parts List for Test

C—Capacitor, 0.01 μF.
R—Resistor, 330 k.
I_C—Hex Scmitt
trigger MC14584 or
equivalent.

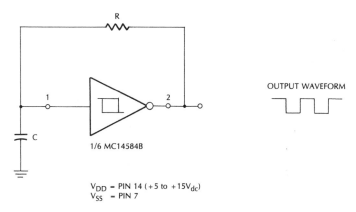

Figure 2-51: Wiring diagram for square-wave generator.

Comments:

With the values shown, you should have an output frequency of about 1 kHz. However, the frequency can easily be altered by selecting different values for C_1 and R_1. *Note:* See Section 3.2 for more about Schmitt trigger ICs.

Wiring and Testing a Simple One-Shot Multivibrator

A basic monostable or one-shot multivibrator (a simple pulse generator) can be made by using the popular CMOS quad 2-input NOR gate, type number CD 4001.

To wire this mono, two NOR logic gates are direct-coupled by connecting the output of one gate to the input of the other, and the output of the second gate is coupled to the input of the first via a simple R-C time-constant network. Figure 2-52 shows the hookup (using one-half of a CD 4001 IC) you can use to wire the mono.

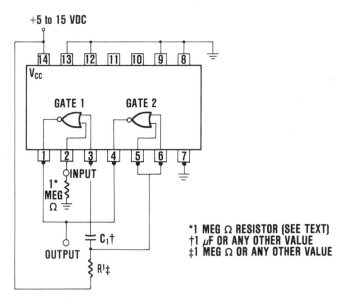

+5 to 15 VDC

Figure 2-52: Wiring diagram for a basic monostable multivibrator (pulse generator). R_1 can be any value from a few k ohms to several megohms. C_1 can be any value from a few pico F to several hundred micro F.

When you wire the IC as shown in Figure 2-52, you are using the first gate (pins 1, 2, and 3) as a NOR logic element, and the second gate (pins 4, 5, and 6) as an inverter, or NOT gate. The other two gates are not used. Therefore, pins 8, 9, 12, and 13 should be at ground potential. *Important:* The input of this circuit (pin 2) must always be connected to ground through a 1-megohm resistor, unless a positive trigger pulse is being applied. An equally important point: *Never* apply an input signal without power on. If these rules are not followed, the IC may be destroyed.

When you have the IC wired as shown in Figure 2-52, and power applied, you should find the input of gate 2 (pins 5 and 6) is at logic high (through R_1), and the output (pin 4) is at a logic low. Also, you should find a logic low at both input terminals of gate 1 (pins 2 and 3). Pin 1 (the output of gate 1) should be at logic high. Next, since both terminals of capacitor C_1 are at logic high, your multimeter will show that the capacitor is completely discharged.

Now, let's see what happens when you use your logic pulser and apply a positive trigger pulse to the input of gate 1 (pin 2). As soon as you apply the input pulse, you should see the output (pin 4 or 1) go high, remain high for a short time, then quickly return back to its normal low state again. Just how long it will stay high depends on what values you chose for R_1 and C_1, i.e., the R-C time constant.

When you applied the input pulse, it drove the output of the first gate to ground potential, which, in turn, dragged the input of gate 2 with it, via the discharged capacitor C_1. This action caused the output of gate 2 to go high, thus holding the output of gate 1 in a low state even when you removed the trigger pulse. However, in time, there was a change of state because, as soon as gate 1 went low as a result of the trigger pulse, C_1 started to charge through R_1 and the capacitor charging voltage was applied to the input of gate 2 through the R_1, C_1 junction.

After a short time (depending on R_1, C_1 values), the capacitor charge voltage rose to the transfer voltage of gate 2 and, at this point, the output of gate 2 switched quickly back into a logic low state. As the output of gate 2 went to logic low, it caused the output of gate 1 to go to logic high. Capacitor C_1 then quickly discharged through the output of gate 1 and the operating sequence was complete.

A final point to note is that since the output pulse can be made longer in duration than the input pulse, this type of circuit is often called a "pulse stretcher." A mono can be used to stretch a brief pulse so that, among other applications, it can be used to drive a relay.

SECTION 2.10: FLIP-FLOP ICs

Bistable or Flip-Flop IC

This integrated circuit is basically a memory device used for storing data in the form of a logic high or low. All flip-flop circuits have two stable states controlled by an input trigger and/or clock pulse. For example, a clocked flip-flop will be triggered (caused to change states) only if the trigger pulse and clock pulse are presented at the same time. Two of the most popular flip-flops in modern digital systems are the delay (or D) and the J-K types. Figure 2-53 A shows a block diagram for a CMOS dual D type and 2-53 B shows a CMOS dual J-K type produced by Motorola.

The D type shown can be used as an edge-clocked flip-flop. In this case it is in a memory mode while the clock input (C, see Figure 2-53) is at logic low, and in memory mode while the clock input is at logic high. The only time the outputs (Q, $\overline{Q}$, see Figure 2-53) can change state is during the brief interval of time it takes the clock signal to make a transition from logic low to logic high or, in some cases, from high to low.

However, there are two other inputs: direct set (S) and direct reset (R). These two inputs influence the outputs without regard to any clocking operations, and their main purpose is to let you set the outputs to either desired logic level before or after a positive-going edge of the clock pulse occurs. This is shown in the truth table for the IC as X = don't care (see Table 2-21).

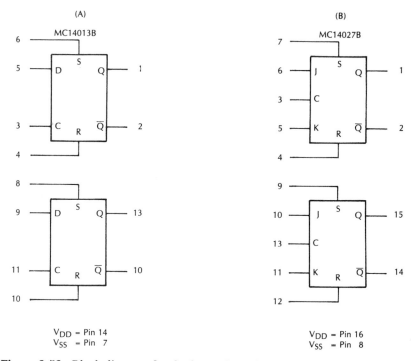

Figure 2-53: Block diagram for dual D and J-K flip-flops. The outputs of these devices are labeled Q and $\overline{Q}$. This notation conforms to modern standards for flip-flops of all kinds. The inputs are always on the left of the drawing—for example, J and K in illustration B.

INPUTS				OUTPUTS		
C*	D	R	S	Q	$\overline{Q}$	
⟋	0	0	0	0	1	
⟋	1	0	0	1	0	
⟍	X	0	0	Q	$\overline{Q}$	NO CHANGE
X	X	1	0	0	1	
X	X	0	1	1	0	
X	X	1	1	1	1	

X = Don't care.
* = Level change.

Table 2-21: Truth table of a type D flip-flop (MC14013B).

Note that the truth table indicates an invalid operating mode, one in which reset = set = 1 (a logic high). Under this particular set of circumstances, the Q and $\overline{Q}$ outputs both go to a logic 1, thus destroying the complemented relationship you should find existing between the two. This D flip-flop is only representative of the modern triggered D flip-flop family now available to the technician in both TTL and CMOS IC packages.

Nevertheless, whether the device is a TTL such as 74247, 74175, 7474, a CMOS 4013 or 4042, or the one just described, basically they all work the same. They may have but a single output (Q), or no preset, but once you understand one, you should have no trouble servicing the others.

Today, it would be next to impossible for an electronics technician to perform his duties without an understanding of the so-called J-K flip-flops. Table 2-22 shows the truth table for the MC14027B dual J-K flip-flop included in Figure 2-53.

INPUTS						OUTPUTS*		
C†	J	K	S	R	Q_n‡	Q_{n+1}	Q_{n+1}	
⟋	1	X	0	0	0	1	0	
⟋	X	0	0	0	1	1	0	
⟋	0	X	0	0	0	0	1	
⟋	X	1	0	0	1	0	1	
⟍	X	X	0	0	X	Q_n	Q_n	NO CHANGE
X	X	X	1	0	X	1	0	
X	X	X	0	1	X	0	1	
X	X	X	1	1	X	1	1	

X = Don't care. † = Level change.
‡ = Present state. * = Next state.

Table 2-22: Truth table for a dual J-K flip-flop MC14027B. A block diagram of this IC is shown in Figure 2-53.

As shown by the truth table, this IC is an edge-triggered flip-flop featuring independent J, K, clock (C), set (S), and reset (R) inputs for each flip-flop. Notice that the multiple inputs provide more control conditions for operating the flip-flop. A logic low on the set input and a logic high on the reset will set the Q outputs to logic Q = Q, $\overline{Q}$ = 1. On the other hand, a 1 on the S and a 0 on the R inputs will reverse the output; i.e. Q = 1 and $\overline{Q}$ = 0.

Basically, you will find that flip-flop ICs are not symbolically designated and can be recognized only by their number and title, as shown by the block diagram in Figure 2-53 A and B. Essentially, the J-K master/slave flip-flop is set with a positive-going edge of a clock pulse, just as is the edge-trigger we just explained. In this case, the first flip-flop (there are at least two in the IC) is called the *master* and is triggered with the positive-going edge of the clock pulse, and the information is transferred to the next flip-flop—called the *slave*—on the negative-going edge of the clock pulse, as shown in Figure 2-54. Figure 2-55 shows the pin configurations and block diagram for a TTL family 7472, J-K master/slave flip-flop that would follow this sequence of operations.

Generally, this is the sequence of operation for a master/slave flip-flop:

1. Isolate the slave from master.

2. Enter information from an internal (within the IC) AND gate's inputs to master.
3. Disable AND gate inputs.
4. Transfer information from master to slave.

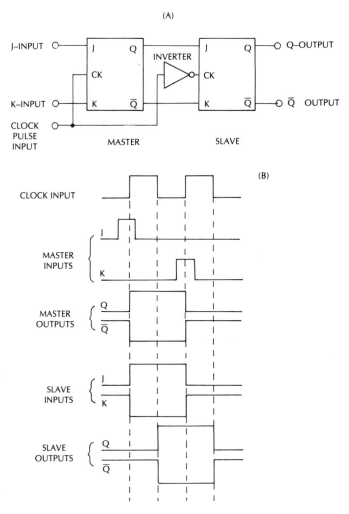

Figure 2-54: J-K master/slave flip-flop. (A) is a logic diagram showing two single flip-flops, master and slave. (B) is a timing chart illustrating inputs, clock pulse, and resulting changes.

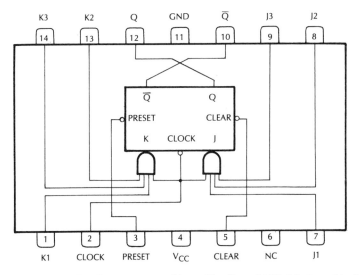

Figure 2-55: TTL family J-K master/slave flip-flop, 7472. Notice, this IC also has a preset and clear. Although these inputs were not shown in Figure 2-54, they are usually found on these types of devices.

Testing the MC14011B as a Touch-Controlled Flip-Flop

Parts List for Test:

I_c—MC14011B
R_1, R_4—Resistors, 100 k.
R_2, R_3—Resistors, 4.7 Megohm.
Touch plate—Any conductive
material cut into four small
pieces (P_1, P_2, P_3, P_4). Each touch
plate consists of two pieces with
a small gap between each piece.

Comments:

To operate the circuit (see Figure 2-56), first touch plates P_3 and P_4 at the same time (with one finger). This should produce a high on the output. Next, touch plates P_1 and P_2 (again, both at the same time), and the output should drop to a low; that is, clear the circuit. *Note:* See Section 3.2 for more about flip-flops.

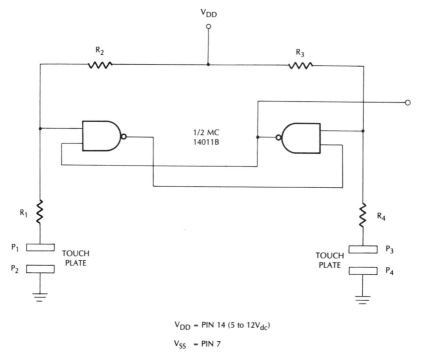

V_{DD} - PIN 14 (5 to 12V_{dc})

V_{SS} - PIN 7

Figure 2-56: Wiring diagram for touch-controlled flip-flop.

SECTION 2.11: DIGITAL COUNTER ICS

In general, readouts used in digital equipment normally display in decimal. However, a decade IC counter automatically resets to 0 after the decimal count of 9, running from 0 to 9, and then repeating. To count to a higher number, the counter output triggers the next largest value. In other words, 09 becomes 10 on the next count, and 199 becomes 200, etc.

Now, what if you want the circuit to start back at 0 at some point before it reaches 9? That may sound like a tough question, but it really isn't. In the following pages, you will see how to do this. However, let's first use the popular and inexpensive 7490 decade counter to learn how a basic counter works.

Actually, this is a monolithic decade counter consisting of four dual-rank, master-slave flip-flops internally interconnected to provide a divide-by-two counter and a divide-by-five counter. It should be remembered that while counters such as the 7490 are properly considered binary counting circuits, it is also possible to view them as frequency dividers. This is because any toggled flip-flop naturally divides its clock frequency.

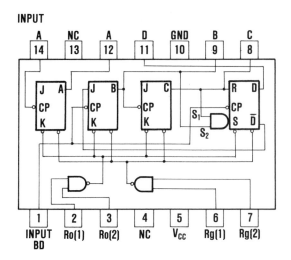

Figure 2-57: Pin configuration for a 7490 decade counter IC.

Figure 2-57 shows the pin configurations of a 7490 decade counter. Notice that the output from flip-flop A is not internally connected to the succeeding stages. For this reason, the count may be separated into three independent count modes: (1) binary coded decimal decade counter, (2) divide-by-ten, and (3) divide-by-two and divide-by-five. For example, the IC can be wired, as shown in Figure 2-58.

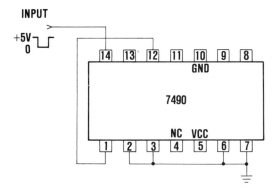

Figure 2-58: How to connect a 7490 IC for a BCD count sequence. *Note:* Output A (Figure 2-57) is connected to input BD for a BCD count.

Referring to Figure 2-57, you will notice that pins, 2, 3, 6, and 7 are all inputs to NAND gates. These pins are grounded for counting. Of course, pin 10 must also be grounded because it is the IC's common ground point. *Note:* It is possible to ground only pins 10, 2, or 3, and either 6 or 7 and the circuit will function as a counter. However, for stable operation, it's best to ground them all, as shown.

Next, refer to Table 2-23. Notice that when there is a logical high (about 5 volts) at output A (pin 12), it represents a count of 1. To put it another way, read under the heading "Decimal Number" the number 1, then read to the right 0001, the binary number 1. Output B (pin 9) is a decimal count of 2, binary number 0010. Output C (pin 8) is a decimal count of 4, binary number 0100, and output D (pin 11) is a decimal count of 8, binary 1000.

BCD COUNT SEQUENCE

DECIMAL	OUTPUT			
NUMBER	D	C	B	A
0	0	0	0	0
1	0	0	0	1
2	0	0	1	0
3	0	0	1	1
4	0	1	0	0
5	0	1	0	1
6	0	1	1	0
7	0	1	1	1
8	1	0	0	0
9	1	0	0	1

Table 2-23: Logic truth table for BCD outputs from a 7490 IC, when connected as shown in Figure 2-58.

If you wire the 7490 IC as shown in Figure 2-58, it should reset to 0 at the next count after 9. Gated direct reset lines are provided (for example, pins 2 and 3 in figure 2-58) to inhibit count inputs and return all inputs to logic 0. The gated circuitry is designed so that when *both* pins 2 and 3 go high, the IC resets to 0. During normal counting, you should find that either pin 2 or 3 (or both) is at a low.

Now, let's see how we can make a counter count to some number other than 10. To see how this can be done, try this. Connect a jumper wire from pin 9 (B, in Figure 2-57) and remove the ground from both pins 2 and 3 (the gated inputs). Incidentally, when you leave a pin ungrounded, it is the same as if you had connected it to a logic high level. Once you have made these changes, you should find that the IC will reset to 0 at the second count; i.e., count from 0 to 1 and back to 0.

If you want the counter to count to 3 (0 to 2), connect pin 2 to pin 9 and pin 3 to pin 12. For a count of 4, connect pin 2 or 3 to pin 8 (don't connect the other gated pin—leave it open). For a count of 5, connect pin 8 to one reset pin and the other reset pin to pin 12. For a count of 6, pins 8 and 9 are connected to the reset pins. If you want to count to a higher number, it is sometimes necessary to make the IC reset by placing pin C (8), B (9), and A (12) at a high all at the same time. The least expensive way to do this is to use a triple AND gate such as the 7411 or a 7408. If you should have to do this, you must trigger the reset terminals with a logic high when all inputs to the AND gate are high. The wiring is: pins 8, 9, and 12 to the AND gate inputs (one pin to each of the three AND gate inputs); connect the AND gate output to the gated pin (pin 2) of the 7490 IC and leave pin 3 of the 7490 disconnected. With this setup, the IC should count 0 to 6 and reset to 0 on the seventh count.

Can you make the 7490 IC do other counts? You sure can. For example, if you want a divide-by-ten count for a frequency synthesizer, or some other application requiring division of a binary count, simply make an external connection (use a jumper wire) between pins 11 (D) and 12 (A). Now, the input count is applied at the BD input (pin 1) and a divide-by-ten square wave is obtained at the output A (pin 12). *Note:* Do not use a jumper wire between pins 2 and 12 (as shown in Figure 2-58) in the divide-by-ten mode of operation. Referring to Figure 2-57, you will notice that there are actually 4 reset inputs, $R_{o[1]}$, $R_{o[2]}$ and $R_{g[1]}$, $R_{g[2]}$. A logic truth table for reset/count is shown in Table 2-24.

RESET/COUNT

RESET INPUTS				OUTPUT			
Ro(1)	Ro(2)	Rg(1)	Rg(2)	D	C	B	A
1	1	0	X	0	0	0	0
1	1	X	0	0	0	0	0
X	X	1	1	1	0	0	1
X	0	X	0	COUNT			
0	X	0	X	COUNT			
0	X	X	0	COUNT			
X	0	0	X	COUNT			

Table 2-24: Logic truth table for the reset/count inputs of a 7490 decade counter IC.

For operation as a divide-by-two, and a divide-by-five counter, you do not have to make any external interconnections. Of course, you must have V_{cc} (min. 4.5V, nom. 5V, and max. 5.5V) on pin 5 and ground at pin 10. Flip-flop A is used as a binary element for the divide-by-two function. The BD input (pin 1) is used to obtain binary divide-by-five operation at the B, C, and D outputs. When you are operating the 7490 in this mode, the two counters operate independently, but all four flip-flops are reset simultaneously.

A few paragraphs ago, we said that if you want to count to a higher number, it might be necessary to use a triple input NAND gate to feed the 7490. But you can encounter divide-by-sixty counters in clock circuits that use only two 7490 ICs and do not need any NAND gates or other such devices. Figure 2-59 shows a wiring diagram that you can use to try this test.

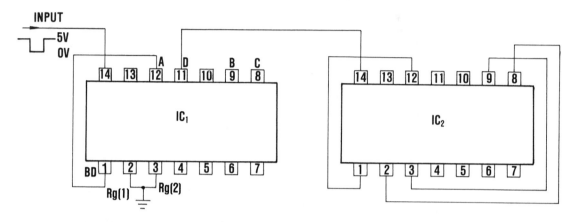

Figure 2-59: Wiring diagram for a divide-by-sixty counter using two 7490 ICs.

IC Binary Counter

When the need for counting events in an electronic system arises, most of us start looking for an IC that is both readily available and low cost. The 7493 4-bit binary counter will satisfy these requirements. The IC shown in Figure 2-60 consists of a group of four master-slave flip-flops that are internally interconnected to provide a divide-by-two and a divide-by-eight

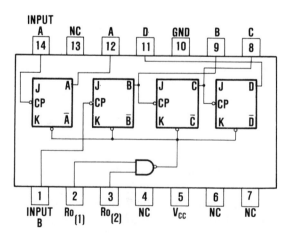

Figure 2-60: Pin configuration for a 7493 4-binary counter.

counter. By examining the pin configuration, you will see that the output from flip-flop A is not internally connected to the succeeding flip-flops. Therefore, the counter can be operated in two independent modes, as a 4-bit ripple-through counter or a 3-bit ripple-through counter, depending on the external connections you use.

If you want the IC to function as a 4-bit ripple-through counter (one pulse ripples along through the flip-flops to provide the outputs), output A must be externally connected to input B. The input count pulses are applied to input A and simultaneous divisions are performed at the outputs, as shown in Table 2-25.

COUNT	OUTPUTS			
	D	C	B	A
0	0	0	0	0
1	0	0	0	1
2	0	0	1	0
3	0	0	1	1
4	0	1	0	0
5	0	1	0	1
6	0	1	1	0
7	0	1	1	1
8	1	0	0	0
9	1	0	0	1
10	1	0	1	0
11	1	0	1	1
12	1	1	0	0
13	1	1	0	1
14	1	1	1	0
15	1	1	1	1

Table 2-25: Truth table for a 7493 IC when used as a 4-bit ripple-through counter.

To illustrate how this counter can be used, let's assume that your needs call for a 4-bit up-counter. This IC is designed for the job. You connect pin 12 to pin 1, pin 5 to V_{CC} (about 5V), then ground pins 10, 2 (R_{o1}), and 3 (R_{o2}), for a count up to 16. On outputs A, B, C, and D, you will have frequency division of 2, 4, 8, and 16. The wiring diagram for this operation is shown in Figure 2-61.

If you want flip-flops B, C, and D to operate as a 3-bit ripple-through counter, you should apply the input pulse to input B. You will find simultaneous frequency divisions of 2, 4, and 8 at outputs B, C, and D. Using this mode of operation permits you to use flip-flop A independently if the reset function occurs at the same time as it does with the reset of the 3-bit

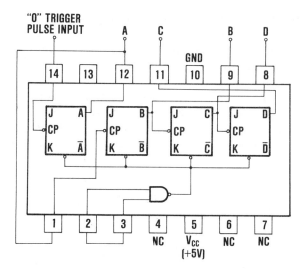

Figure 2-61: Wiring diagram for connecting a 7493 IC for use as a 4-bit binary counter. Supply voltage (V_{CC}) min. 4.75, nominal 5, max. 5.25. The minimum voltage required to insure logical 1 at any input terminal is a 2 V at min. V_{CC}. Max. input voltage for logical 0 at min. V_{CC} is 0.8 V.

ripple-through counter. The gated direct reset line (pins 2 and 3) is provided to stop the count and simlutaneously return all four flip-flop outputs back to logical 0.

Using and Testing Binary Up/Down Counters

Comments:

You can use a 74193 4-bit binary up/down counter to count up from 0 to 15 or, by using other inputs, count down from 15 to 0. As shown in Figure 2-62, this up/down counter has an input for count down (pin 4) and another for count up (pin 5).

The outputs of the four master-slave flip-flops contained in the IC are triggered by a low-to-high transition of either clock input (pin 4 or 5). Synchronous operation is possible by having all flip-flops clocked simultaneously so that the outputs change state at the same time, when instructed to do so by the steering logic. However, the outputs may also be preset to any state by entering the desired data at the data inputs while the load input is low. In this case, the output will change to agree with what you have entered independent of the count pulses. This feature allows counters to be used as programmable dividers by simply changing the count length with the preset inputs.

The two outputs, borrow (pin 13) and carry (pin 12), can be connected to clock-down and clock-up inputs of a subsequent counter that you need to use when working with numbers greater than 15. To put it another way, the 74193

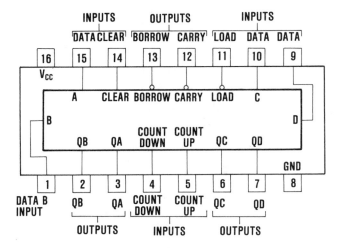

Figure 2-62: Pin configurations for a 74193 synchronous 4-bit binary up/down counter with preset inputs.

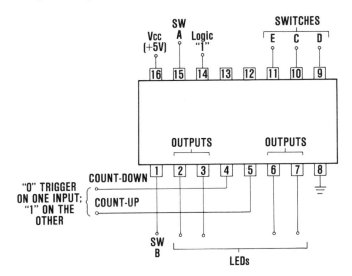

Figure 2-63: Up/down binary counter: testing a 74193.

counter IC is designed to be cascaded without the need for the external components. Simply connect pins 13 and 14, as explained, and you can cascade both the up and down functions for higher number capabilties. The basic circuit for this test is shown in Figure 2-63.

Procedure:

Step 1. Wire in the 5 switches (A, B, C, D and E).

Step 2. Connect LEDs to output pins 2, 3, 6, and 7

Step 3. Insert the IC into a socket.

Step 4. Connect $+V_{cc}$ to pin 16, ground pin 8. Set V_{cc} to 5 volts.

Step 5. Data pins (switches A, B, C and D) are set high for the desired number to be entered (see Figure 2-63).

Step 6. Load pin (switch E) is normally high, but you must set it to *low* momentarily in order to load the IC (see Figure 2-63).

Step 7. To count up, count-down input must be at a high. To count down, count-up input must be high.

Step 8. Pulse either pin 4 (count down) or pin 5 (count up), depending on which way you want to count.

Step 9. Place a logic 1 level pulse on pin 14, to clear the IC (see Figure 2-63).

Typical clear, load, and count sequences are shown in Figure 2-64.

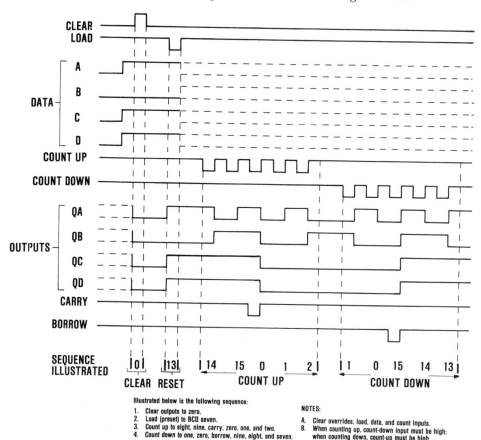

Figure 2-64: Typical clear, load, and count sequences for the binary counter 74193.

SECTION 2.12: SHIFT REGISTER ICS

You may encounter several different families of shift register. For example, there are the CMOS 4-bit shift register (MC140358), the Schottky TTL 4-bit register (SN74LS194A), and the standard TTL 4-bit shift register (7495). But, they are all constructed using inverters, gates, and flip-flops.

In keeping with the theme of this book—keeping things as simple as possible—we will use the 7495 4-bit right-shift, left-shift register to explain the digital circuits used to perform most of the basic operations required of a register.

A series register is a group of flip-flops used for temporary storage, and the number of flip-flops determines the amount of data per unit. Therefore, it follows that the 7495 4-bit shift register is constructed using four flip-flops. The circuit layout consists of four R-S master-slave flip-flops, four AND-OR invert gates, and six inverters configured to form a register that will perform right-shift, left-shift, or parallel-out operations, depending on what level is used on the input to the mode control.

The mode control input (pin 6) is located in the upper left-hand corner of the logic diagram (Figure 2-65). Right-shift operations are performed when you apply a logic level 0 to this mode control pin. Serial data is entered at the serial input (pin 1) and shifted one position right on each clock 1 pulse. When you are operating the register in the serial mode, clock 2 (pin 8) and

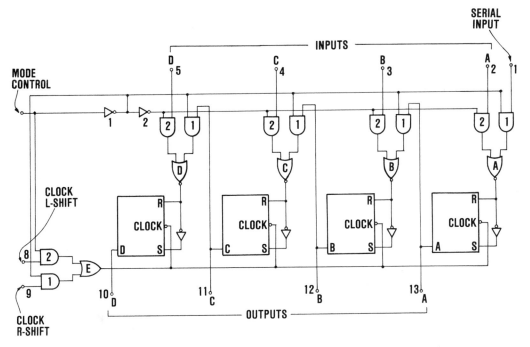

Figure 2-65: Logic diagram for a 7495 4-bit right-shift, left-shift register.

parallel inputs A through D (pins, 2, 3, 4, and 5) are inhibited. Each flip-flop and the AND-OR invert combinations connected to the R-S inputs are shown in Figure 2-65.

Data can be entered at the parallel inputs (pin 2, 3, 4, and 5) when a logical 1 level is applied to the mode control. The data entered at the parallel inputs is transferred to the data outputs (pins 10, 11, 12, and 13) on each clock 2 (pin 8) pulse. When you are operating in this mode, you can perform shift-left operations by externally tying the output of each flip-flop to the parallel input of the previous flip-flop (up to flip-flop C, with serial data entry at input D). *Note:* Information must be present at the R-S inputs prior to clocking, and transfer of data occurs on the falling edge of the clock pulse.

Now, let's say you are operating the IC in a shift-right mode, serial input/serial output. That is, the output of one flip-flop feeds into the inputs of the next. Table 2-26 shows the serial loading and transfer action of the register. This table shows the loading sequence and that the register is completely loaded on the fourth clock pulse. On the last clock pulse, the register contains the binary number 1010 in memory. It should be obvious that if you continue applying pulses, the data (binary number 1010) you have in the register will shift right out of the IC. This means that the data will be completely lost unless you connect to another register or some other type of digital circuit.

CLOCK 1 PULSE (R-SHIFT)	SERIAL INPUT	FLIP-FLOP 1	FLIP-FLOP 2	FLIP-FLOP 3	FLIP-FLOP 4
1	0	0	0	0	0
2	1	1	0	0	0
3	0	0	1	0	0
4	1	1	0	1	0

Table 2-26: 4-bit shift register operating in a right-shift mode.

Wiring and Testing a 4-Bit Right-Shift, Left-Shift Register

Comments:

In this test, two TTL ICs, a 7407 and a 555 timer, will be used with a 74194 4-bit bidirectional shift register. Figure 2-66 shows how the 74194 can be breadboarded to a set of control switches and to the 555 timer for a clock signal.

The clock pulse is manually generated each time you depress the clock switch. When you activate the switch, you should find a *positive* pulse at the output (pin 3) of the 555 IC. But note, pin 10, the output of the inverter, should have an inverted pulse going into the clock input of the 74194 IC.

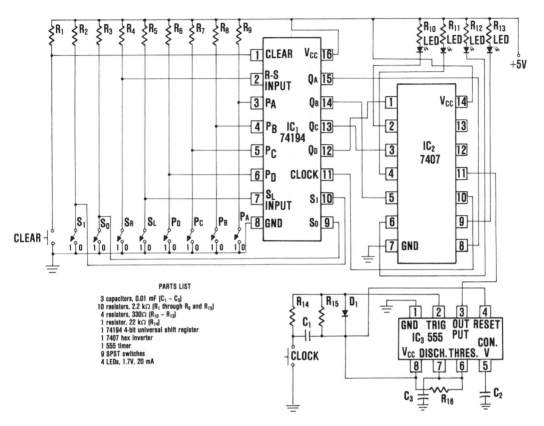

Figure 2-66: A 4-bit bidirectional universal shift register (74194) wiring diagram, where IC_2 is the switch-debouncing circuit and IC_3 is a monostable multivibrator clock circuit.

The 555 timer circuit has an 11 millisecond delay, which insures that only one pulse will be delivered to the shift register each time you activate the clock switch, even though there may be several bounces of the switch contacts. For this test, you do not have to worry about debouncing the other switches because you can get away with some switch contact bounce.

As shown (all switches open), the inputs are all at a logical high. When you close any of the switches, you are setting that input to a logical low. The outputs from the shift register are active-high, but to use LEDs on the outputs, you need active-low logic, which is why you must use the inverter stages of the 7404 hex inverter on the outputs of the 74194 register. If you are checking the circuit, you should find that the LEDs will light up whenever the outputs (QA, QB, QC, and QD) of the 74194 are at a logical high. The register has four distinct modes of operations. These are shown in Table 2-27.

	MODE CONTROL	
	S_1	S_0
PARALLEL (BROADSIDE) LOAD	H	H
SHIFT RIGHT (IN THE DIRECTION QA TOWARD QD)	L	H
SHIFT LEFT (IN THE DIRECTION QD TOWARD QA)	H	L
INHIBIT CLOCK (HOLD)	L	L

Table 2-27: Four modes of operation that you can utilize when working with the 74194 shift register.

Procedure (parallel in/out mode):

Step 1. Clear the register by depressing the clear switch.

Step 2. As shown in Table 2-27, you must set S_0 and S_1 switches to a logical high.

Step 3. Enter a 4-bit word (0101, or whatever you wish) by setting the switches on the parallel input pins (pins 3, 4, 5, and 6) of the register.

Step 4. Now, because in the parallel mode, data is loaded into the associated flip-flop and appears at the outputs only after a positive transition of the clock input, you will have to momentarily depress the clock switch. When you do this, you should see whatever 4-bit word you entered in Step 3 appear at the outputs.

Step 5. Check the outputs (LEDs). There should be no change. The purpose of this step is to prove to yourself that the register will hold the first-entered data until you again depress the clock switch. In other words, the memory circuit will hold the last information entered until clocked again. *Note:* It is not necessary to clear the register before clocking in a new word. You should find that the circuit will write over the 4-bit word you previously entered. Also, during the loading, the serial data flow is inhibited.

Procedure (Right Shift):

Step 1. As shown in Table 2-27, set S_1 to low, S_0 to high.

Step 2. Depress the clear switch. This should set all LEDs to 0s.

Step 3. Set switch SR_1 (to pin 2 of the register IC) to either high or low.

Step 4. Activate the clear switch *one time* to enter that bit (see Step 3) into the QA section of the register IC.

Step 5. Enter one more bit at pin 2 (SR_1). Again activate the clear switch one time.

Step 6. Repeat Step 5 two more times, i.e., until you have four bits of data entered into the register.

Step 7. After you have completed the preceding six steps, you should have a 4-bit word entered and stored in the register, with the bit you first entered shown in the LED connected to the QD output of the register.

Procedure (Left Shift):

Step 1. Refer to Table 2-27 and set mode control S_1 to high, S_0 to low.

Step 2. Now, simply repeat each step given in the procedure for the shift-right *except* you must now enter the new data in the shift-left serial input (pin 7) by using switch SL_1.

Step 3. Finally, you will find that it is impossible (or it should be) to clock the flip-flops in the register if both mode control inputs are at a logic 0 (switch closed). Incidentally, if you are having trouble with the test, check the mode control switches to see that they are not both connected to ground (not set to logical 0).

SECTION 2.13: MEMORY ICs

As seen in previous sections, a flip-flop is a basic memory device. There are, of course, many other types of memory circuits used to store binary bits that represent data, but basically, there are two: *Random Access Memory* (RAM) and *Read Only Memory* (ROM). A RAM contains data that may be electrically removed at any time new data is stored in its place. The other type—ROM— has data (computer programs, etc.) fixed in the memory and uses several different techniques to store the permanent pattern.

You will find that memory ICs are manufactured using TTL, CMOS, and ECL techniques. However, in all cases they are built around a system called *memory cells.* Each memory cell is only one of many identical devices in a memory IC. Now it stands to reason that if a memory IC contains many storage elements, there must be some provision that will let the computer, electronic instrument, etc., select one particular memory cell at a time. Memory ICs have a *memory cell select* and there are provisions for choosing either a writing (store) or a reading operation. For example, the MCM145101, a 256 × 4-bit static RAM, is a random access memory having 1,024 separate memory cells, that can accept, store, and read out 256 different 4-bit words. Figure 2-67 shows the block diagram for this CMOS memory IC.

Since this memory has 256 word-storage locations, it follows that it must have a cell array that adds up to 1,024. Refer to Figure 2-67 and you'll see that this IC does—32 rows, 32 columns. The address inputs (cell selection by row and column) are A0 through A7. These address inputs are processed in the row and column decoders which, in turn, select the correct cell for information storage.

There is a control terminal that determines whether data will be written into memory via the data inputs (DI) or read out via the data outputs (DO),

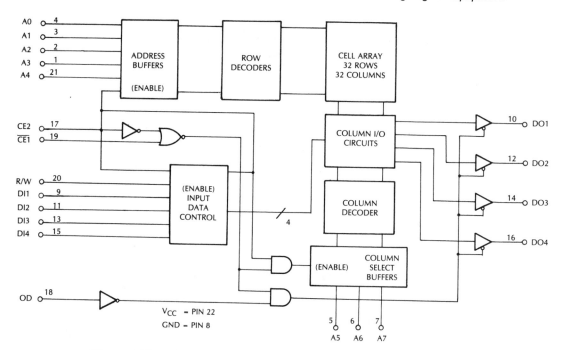

Figure 2-67: The MCM145101 256 by 4-bit memory IC block diagram.

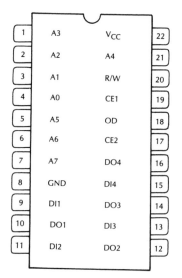

CE1	CE2	OD	R/W	D_{in}	OUTPUT	MODE
H	X	X	X	X	HIGH Z	NOT SELECTED
X	L	X	X	X	HIGH Z	NOT SELECTED
X	X	H	H	X	HIGH Z	OUTPUT DISABLED
L	H	H	L	X	HIGH Z	WRITE
L	H	L	L	X	D_{in}	WRITE
L	H	L	H	X	D_{out}	READ

Figure 2-68: Pin configurations and truth table for a MCM145101 256 × 4-bit static RAM.

plus other control operations. Writing a 4-bit word into memory is a matter of setting each of the IC's pins to the proper logic level. See the truth table and pin assignment in Figure 2-68.

As we have shown, the primary feature of a RAM IC is that it is possible to read and write data with it. A ROM, however, is capable of performing only read-out operations. Forgetting erasable programmable read only memories. (EPROMs) for the moment, the programming of a ROM IC is internal and permanent. The MCM14524 device is a CMOS large-scale integration (LSI) ROM organized with a 256 × 4-bit pattern. It is, in other words, a 1024-bit ROM. The internal ROM programming is specified by the user, but it is the manufacturer's (Motorola, in this case) responsibility to implement the actual programming of the IC during the last phase of the manufacturing process. Programming of an IC such as this example is usually a custom job. ROM ICs found in personal computers and video games are an example of this type of custom programming.

The advent of programmable read only memories (PROMs) and erasable programmable read only memories has been a boon to the electronics industry. To program a PROM, you blow selected fuses that are in each memory cell. A new IC of this type will have all its on-chip memory cell fuses intact and the outputs will be 1s or 0s, depending on how it was manufactured. You blow selected fuses electrically with external power supplies. Blowing the fuse of each memory cell may place that cell at a logic high, whereas in other PROMs it could represent a permanently stored logic low.

The PROM is usually considered to have permanent data stored in its memory. EPROMs, as the name suggests, are erasable. One type has a quartz window on top which is transparent to ultraviolet light. You erase the memory's contained data simply by exposing the window to an external source of ultraviolet light. As shown in Figure 2-69, the transparent lid is easily placed under the ultraviolet light. However, because this light is harmful to eyes, the light and EPROM are always placed inside a closed case. The erasure time is usually 15 to 20 minutes when using a lamp with 12,000 $\mu W/cm^2$ power rating. Another IC that contains an EPROM of this type is the microcomputer MC68701.

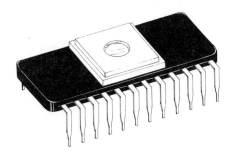

Figure 2-69: 1024 × 8 alterable (ultraviolet) ROM.

A different type, an MOS device, is called a 16×16-bit electrically erasable programmable read only memory. This EPROM, the MCM2801, offers in-system erase and reprogram capability. Figure 2-70 shows a pin assignment diagram and pin names for this IC.

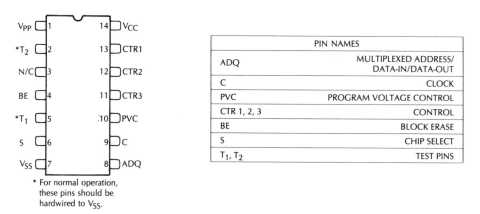

PIN NAMES	
ADQ	MULTIPLEXED ADDRESS/ DATA-IN/DATA-OUT
C	CLOCK
PVC	PROGRAM VOLTAGE CONTROL
CTR 1, 2, 3	CONTROL
BE	BLOCK ERASE
S	CHIP SELECT
T_1, T_2	TEST PINS

* For normal operation, these pins should be hardwired to V_{SS}.

Figure 2-70: 16×16-bit serial electrically erasable PROM.

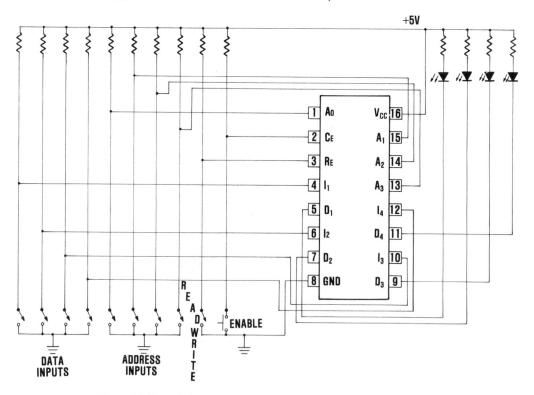

Figure 2-71: Wiring diagram for testing the 7489 TTL RAM.

Testing a Read/Write Memory IC

Comments:

Figure 2-71 shows a simple circuit you can build to test a 7489 64-bit read/write memory IC. As with most digital IC test circuits, as long as you leave the data and address switches open, you are placing a logic 1 on their respective input terminals. Or, when you close one of the switches, you are placing a logic 0 on the respective input terminal.

All ten resistors that are connected to the switches are 2.2 kΩ. The other four (those connected to the LEDs) are usually 330Ω.

The memory read-and-write toggle switch is used to place the IC into either the read or write mode of operation. When you leave this switch open, you are placing a logic high on pin 3 (placing the IC in the read mode). Closing the switch places a logic low on pin 3 (changes the operation to a write mode). The push-button switch (memory enable) must be depressed before either the read or write operation can be performed, but *not while changing the address.* If you do not follow these directions, you are sure to end up with some confusing data on the outputs that has no resemblance to your truth table outputs. Table 2-28 is a truth table for a 7489 RAM.

OPER-ATION	ADDRESS LOCATION	ME PIN 2	R/W PIN 3	OUTPUTS $D_2 - D_4$
READ	$A_0 - A_3$	0	1	0's FOR STORED DATA
WRITE	$A_0 - A_3$	0	0	ALL 1's

Table 2-28: Truth table for a 7489 RAM.

Procedure:

Step 1. Apply V_{CC} to pin 16, ground to pin 8. Remember that this IC is a volatile memory. Do not remove power during this or the next test procedure.

Step 2. Write a logic 0 at all address switches.

Step 3. Read all locations. You should read a logic low at all of them. If you don't, the IC is probably defective.

Step 4. Write a logic 1 at all address switches. You should read a logic high at all locations during this part of the test. If not, the IC is probably defective.

Testing All 16 Locations of the 7489 RAM

Procedure:

Step 1. Make up a truth table for the binary data you wish to store at each of the 16 locations.

Step 2. Select the desired address by setting the address switches to the proper position.

Step 3. Be sure you place the switch on pin 2 (enable) to a logic low.

Step 4. Switch the read/write switch (on pin 3) to a logic 1 and then back to logic 0. The LEDs should blink (turn off and on) indicating the binary data you entered is stored at the selected address.

Step 5. Repeat the preceding steps for the other 15 address locations. *Note:* Remember, the 7489 output is the *complement*. To increase the word length, you can use two or more 7489s. As an example, Figure 2-72 shows the basic hookup for using two of these ICs.

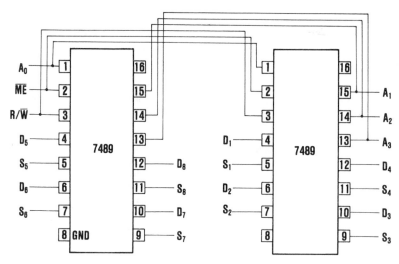

Figure 2-72: Wiring diagram for using two 7489 ICs to increase word length (32×4).

The address and read/write operations require a few more steps than when using a single IC. They are:

Step 1. To read the first 16 locations, address A0 to A3, place $\overline{ME}$—1 high, $\overline{ME}$—2 low, to enable IC1. You should see the data stored on pins 5, 7, 9, and 11 of the second IC, or at the sense outputs.

Step 2. To read the second 16 locations, address inputs A4 through A7, place $\overline{ME}$—1 high and $\overline{ME}$—2 low, to enable IC2. You should see the outputs at pins 5, 7, 9, and 11 of the second IC, or at the sense outputs.

Step 3. The write operation: Select the proper pin 2 input, address, and the binary data you wish to be stored. Place this data at inputs D_1 through D_4 when you set pin 3 to a logic low. Although the wiring diagram shows the address pins of each IC as separate inputs, they actually can be tied together because only one IC is operational at a time, i.e., when its $\overline{ME}$ input is placed on a logic low. *Caution:* Be sure the IC you are not loading has its $\overline{\text{ME}}$ input at a logic high (open).

One thing you should watch when you select IC2, the memory location 17 is address 0000 and, when you program the following locations, they are addressed respectively up to locations 32. This locations' address should be 1111.

Programming a PROM with a Truth Table

Comments:

Techniques for programming PROMs differ according to the technologies used to implement the device. Certain types of MOS PROMs (EPROMs, for example) can be erased and reprogrammed, but bipolar TTL PROMs can be programmed only once. Now, Signetics' 8223 normally comes with all outputs set at a logic 0, but with some PROMs, burning the internal fuse open will represent a stored 0, whereas in others (such as the 8223), it will represent a stored 1. Because of this, it is very important that you have

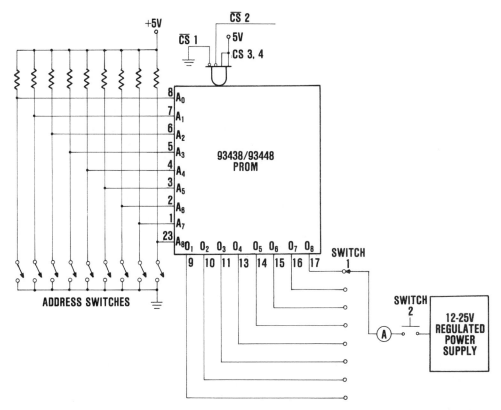

Figure 2-73: Example of a PROM programming circuit.

program instructions, specifications sheets, etc. for the PROM you are working with.

If you are programming a PROM from scratch, a truth table must first be drawn up. You will also use this truth table when the stored information is being read. The address code depends on the equipment in which the PROM is used (a computer or whatever). As an example, a partial truth table that might be used to program the PROM shown in Figure 2-73 is shown in Table 2-29.

PROGRAM SEQUENCE	ADDRESS INPUTS									OUTPUTS								
	A_8	A_7	A_6	A_5	A_4	A_3	A_2	A_1	A_0	O_8	O_7	O_6	O_5	O_4	O_3	O_2	O_1	O_0
1	0	0	0	0	0	0	0	0	0	0		0	0	0	0		1	1
2	0	0	0	0	0	0	0	0	1	0	0	0	0	0	0	1	0	1
3	0	0	0	0	0	0	0	1	0	0	0	0	0	0	1	1	1	
4		0	0	0	0		0	1	1	0	0	0	0	1	0	1	0	
•	•	•	•	•	•	•	•	•	•	•	•	•	•	•	•	•	•	
•	•	•	•	•	•	•	•	•	•	•	•	•	•	•	•	•	•	
•	•	•	•	•	•	•	•	•	•	•	•	•	•	•	•	•	•	

Table 2-29: A partial truth table that might be used to program a PROM with the setup shown in Figure 2-73.

Again, it must be emphasized that this is only an example of how the programming of a PROM is accomplished and you probably will have to refer to the manufacturer's data book, spec sheets, or some other such instructions for the specific PROM IC you want to program. A program setup is illustrated in Figure 2-73. Logic switches are used to select the proper address. In general, these switches do not need to be debounced. You must also apply a programming pulse to the output pin associated with the bit to be programmed. More than likely, you'll find that the other outputs may be left open or tied to any high. *Note:* In almost every case, when using manually operated switches to program one bit at a time, you should not try to program more than one output at a time.

Procedure:

Step 1. Apply V_{cc} and ground to the IC.

Step 2. Use the address switches to select the desired address. The example truth table shows all switches are to be set at logic 0.

Step 3. Set SW1 to output 0.

Step 4. Depress SW2 for a very short time (less than 1/2 second, if possible) and release. Watch the current meter to see that you do not exceed the manufacturer's recommended value. This step is to provide the amount of current necessary to permanently open a nichrome fuse link (other materials are used with these fuse links, however, the nichrome is very popular).

Step 5. When you burn the fuse link open to produce a logic 0 or logic 1, depending on the circuit design, a considerable amount of on-chip heat is developed. Therefore, in all cases, allow several seconds for the chip to cool down before you bring the readout pin to the proper level for a verify check.

Step 6. Set SW1 to output 02 and repeat Steps 4 and 5.

Step 7. Now, referring to Table 2-29 (program sequence 1), we see that the word 11000000 is now programmed in address A_o through A_8. You should continue to select each address and program each word (one bit at a time) by burning open each of the fuses according to your truth table, until you have completed programming the PROM IC.

Programming Erasable PROMs

The erasable PROM (EPROM) is a chip that the electronics technician should become familiar with because these very useful devices are being increasingly accepted as a circuit element. For example, Ultra-Violet Products, Inc., San Gabriel, CA 91778, is selling an EPROM erasing lamp that is low-cost and designed specifically for the small-system user, computer hobbyist, and experimenter. This high-powered unit will erase up to four chips at one time in as little as 20 minutes and it's easy to use. Just plug into the nearest outlet, load with chips and go! The entire unit weighs only one pound. *Warning:* Short-wave ultraviolet light can cause damage (the same as electric arc welding or sunburning) to your eyes and skin. Do not look directly into an EPROM erasing lamp. Also, avoid shining an ultraviolet lamp on reflective surfaces. Special safety goggles are available from such companies as Ultra-Violet Products, Inc.

Just as there are closed or fused-open junctions in a PROM array, the EPROM uses static charges on MOSFET transistors to achieve the effects of an open or closed junction. As another example, one type of EPROM made by National Semiconductor is the 2048-bit MM5203, packaged in a 16-pin DIP with a quartz window on top. As before, the quartz window is transparent to short-wave ultraviolet light (2530 Angstrom units). To erase the unwanted data, you simply expose the window to an ultraviolet device, as has been explained.

SECTION 2.14: SERVICING TECHNIQUES FOR DIGITAL CIRCUITRY

Using a Multimeter for Digital Circuit Testing

Testing digital integrated circuits requires you to adopt a new workbench philosophy, if all your testing has been restricted to analog circuits.

Traditionally, when the individual components of a circuit were accessible, you could use signal generators, diode testers, transistor testers, oscilloscopes and the like to test almost anything electronic. Not today—especially when testing digital integrated circuits. In fact, about the only traditional test gear that can be used when working with digital circuits are the multimeter and oscilloscope.

With the multimeter, you can check the power supply voltage of a malfunctioning digital circuit, and other so-called *dc characteristics*. As a practical example, let's assume that you want to check the dc characteristics of the familiar 7400 IC. This IC is a quadruple 2-input positive NAND gate. This gate, like all digital circuits, is simply an ON-OFF switch. ON is represented by a logic level 1 voltage, and OFF by a logic level 0 voltage. The most important question at this point is, "What voltage would you actually measure when a logic level 1 is measured at a certain pin?" To answer questions like this one, we must refer to the IC specifications sheet and study the IC's electrical characteristics. Table 2-30 shows the electrical characteristics of a 7400 IC gate.

PARAMETER		TEST CONDITIONS*	MIN	TYP**	MAX	UNIT
$V_{in(1)}$	Logical 1 input voltage required at both input terminals to ensure logical 0 level at output	V_{CC} = MIN	2			V
$V_{in(0)}$	Logical 0 input voltage required at either input terminal to ensure logical 1 level at output	V_{CC} = MIN			.08	V
$V_{out(1)}$	Logical 1 output voltage	V_{CC} = MIN I_{load} = −400 μA V_{IN} = 0.8V	2.3	3.3		V
$V_{out(0)}$	Logical 0 output voltage	V_{CC} = MIN V_{in} = 2V I_{sink} = 16 mA		0.22	0.4	V
$I_{in(0)}$	Logical 0 level input current (each input)	V_{CC} = MAX V_{in} = 0.4V			−1.6	mA
$I_{in(1)}$	Logical 1 level input current (each input)	V_{CC} = MAX V_{in} = 2.4V V_{CC} = MAX V_{in} = 5.5V			40 1	μA mA
I_{os}	Short circuit output current	V_{CC} = MAX	−18		−55	mA
$I_{CC(0)}$	Logical 0 level supply current	V_{CC} = MAX V_{in} = 5V		12	22	mA
$I_{CC(1)}$	Logical 1 level supply current	V_{CC} = MAX V_{in} = 0		4	8	mA

*For conditions shown as MIN or MAX: Minimum = 4.75V, Maximum = 5.25V
**All typical values are at V_{CC} = 5V, TA = 25°C

Table 2-30: Electrical characteristics of a 7400 IC needed for testing the device.

You'll notice (Table 2-30) that some of the parameters are gate input currents (I_{in}). In general, gate inputs, except for those where CMOS devices are included, require a current supply, or deliver current to a driver stage or other circuit. If the circuits you are working with involve several gates fanned in or fanned out, it's especially important that you take the current parameters into consideration. Another current parameter that you need to watch is *short-circuit output current* (I_{os}). This parameter is an indication of the ability of the IC to withstand a short circuit. *Note:* As a general rule, you should not short more than one internal IC circuit during any test.

In some cases, the minimums are the most significant values and, in others, the maximums. The other value, typical (TYP), should not be used as anything except to get an idea how the device should perform during a test run under the specified conditions. To make a logic 0 output, V_{in} (1), test of a gate in the 7400, you can use the test setup shown in Figure 2-74.

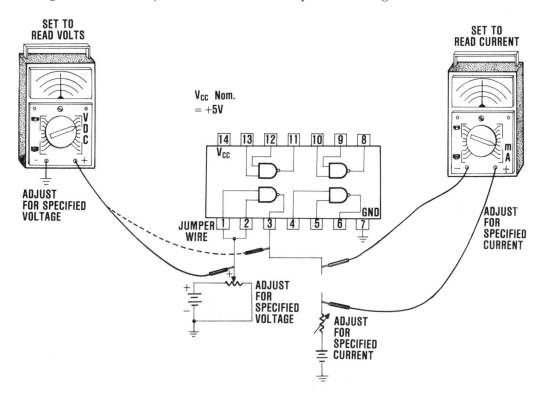

Figure 2-74: Test setup for checking a 7400 gate logic 1 input level, logic 0 output, and specified current.

The second parameter shown in Table 2-30 can be used to test for the opposite condition, i.e., V_{in}, logical 1 level on the output. When you make this test, be sure to *reverse your current meter leads*. Notice that the current meter connections are the exact opposite in Figure 2-74 from those shown in Figure 2-75 for this test. The I_{cc} (0) and I_{cc} (1) typical values given for this IC are about standard for TTL logic.

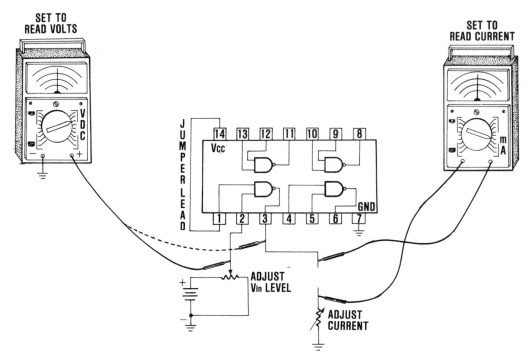

Figure 2-75: Test setup for checking V_{in} (0) parameter shown in Table 2-30; i.e., to insure logical 1 level at output.

Figure 2-76 shows a test setup that you can use to check the input current under I_{in} (0) conditions. You should check each input pin of the IC separately if you are testing the 7400, or any other TTL gate for that matter.

You may find that different manufacturers of the 7400 IC list lower or higher current values than shown in Table 2-30 for I_{in} (0) and I_{in} (1), but, in most cases, this is not of major importance. What you are mainly interested in when testing TTL devices such as the 7400, is that each of the inputs has about the same value of input current during testing. If they don't, it's almost certain that you have a bad IC.

The next parameter of interest (shown in Table 2-30) that you may wish to check is I_{in} (1). Again, you should check each gate input separately. Also, don't forget to check your ammeter for the correct connections, as shown in Figure 2-77. The same rule applies when checking the I_{in} parameter. If all inputs do not have somewhere close to the same current reading, you can suspect a defective IC.

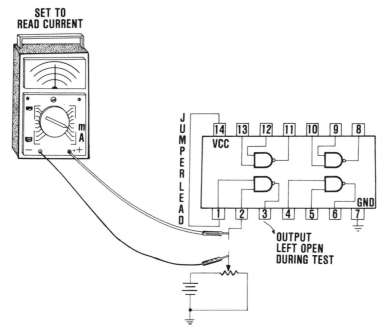

Figure 2-76: I_{in} (0) test setup for each input of a 7400 IC, or similar TTL logic.

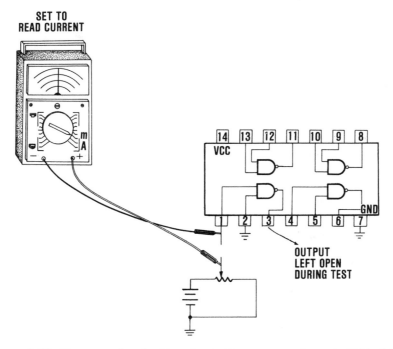

Figure 2-77: Test setup for checking the I_{in} (1) parameter shown in Table 2-30, or any similar TTL device.

Referring to Table 2-30, you'll notice I_{os} follows I_{in} parameters. The test setup for checking the short-circuit output current is shown in Figure 2-78. There are three cautions to be observed during this test:

1. *Not more than one output should be shorted at a time.*
2. *A current limited power supply should be used during the test.*
3. *The IC must be disconnected from any original circuits before performing any of these current tests.*

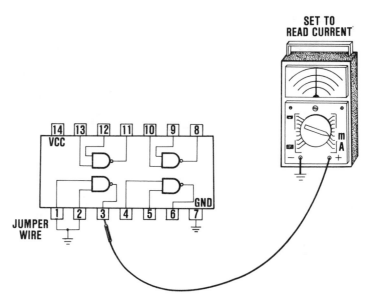

Figure 2-78: Test setup for checking short circuit output current (I_{os}). *Do not make this test without reading cautions listed in text.*

The final parameter (I_{cc}) may be tested for logical 0 level supply and logical 1 level supply (all gates to be in the same condition). Sometimes you will find a spec sheet that will list the current for each gate. In this case, the total current (I_{cc}) will be the sum of the rated current per gate. See Figure 2-79 for test setup.

How to Test with a Logic Probe

One test instrument (others are explained in Section 2.5) that almost anyone who works with digital circuits has become familiar with is the *logic probe*. The instrument is hand-held and has one or more indicator lights near its probe tip. An example is shown in Figure 2-80. The logic probe shown also includes a memory indicator and memory reset switch. The purpose of the memory is to make it possible for a change in logic level to be observed. If you are working with a single indicator light probe, a lit indicator is a logic 1 and

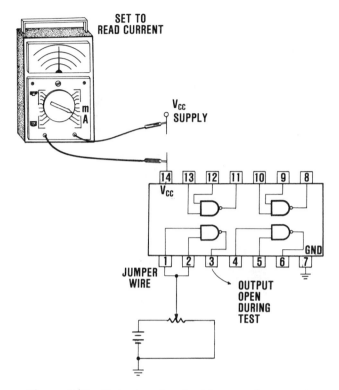

Figure 2-79: Test setup for checking supply current.

an unlit indicator is a logic 0. The probe in Figure 2-80 has two logic indicators; a light for the logic 1 state, and another for the logic 0 state.

To use a logic probe, connect the two wires that are provided for an external connection to the power supply (generally, V_{cc} and ground; the red wire is positive and the black is negative). When testing with the probe, simply touch a PC card run or the pin of an IC and observe the indicators, or indicator, whichever the case may be.

Because a probe works best when observing low-repetition rate pulses of short duration, you should use the probe in either of the following two ways. One, you can start the circuit under test operating at its normal clock rate and then check only key signal lines; or, two, if the clock rate is too fast to be observed effectively, slow the system clock rate down by using a low-frequency clock source until you can observe the state using the probe.

I would like to say, "That's all there is to it." Unfortunately, that *isn't* all there is to it because, when you slow the clock down, you'll probably find that the circuit, due to propagation delay, no longer reacts the same. Many prototype problems that occur are due to the delaying of parallel signals, extraneous voltage pulses produced by propagation delays in two or more logic paths, or because of PC board inductance and capacitance (both of these

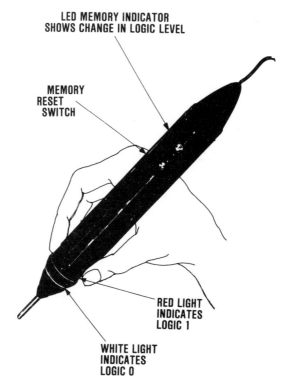

Figure 2-80: A probe with two logic indicator lights near its tip.

are frequency sensitive). If you run into these problems, it's possible that you may have to use an instrument such as a logic comparator or analyzer rather than a logic probe.

Testing instructions, particularly those for digital circuits, may seem very loose if you are familiar with electronic component specs in general. When checking logic levels, it's important to remember that *absolute amplitudes are unimportant.* Logical states are defined by threshold levels of the logic family you're checking. For example, for TTL the low threshold is 0.4 volts and the high threshold is 2.4 volts. When the amplitude of the signal you are checking is *less* than 0.4 volts, it is considered to be at logic 0 level and the logic probe light that indicates logic 0 level should light (see Figure 2-80). When the TTL signal under observation is above 2.4 volts, it is at a logic 1 level and the upper light at the probe tip should light. However, some probes will instantly recognize high, low, or intermediate levels, open circuits, and pulsing modes.

In summary, the low-cost logic probe works best at low-frequency levels but can be used on high-frequency circuits, if you restrict your tests to checking the key control lines of the system. (Incidentally, there are high-speed logic probes that capture pulses as short as 10 ns, but they cost a bit more.) Also, if your logic probe has a memory, you aren't left defenseless when trying to cope with 50 ns pulses, which are more than long enough to trigger IC logic devices. This can be a problem if you do not have a high-speed triggered sweep oscilloscope. A logic probe can overcome the oscilloscope problem.

How to Recognize IC Failure Modes

A solid grasp of the contents of this section will help you in several ways. It will make your job easier, speed up your work, and may increase your earning power. Furthermore, many times your servicing jobs will have a few self-created bugs, and any system can malfunction due to a defective component. Unfortunately , troubles like these are generally hard to identify and track down. You must know what the output of a digital circuit (unlike many other solid state circuits) will be for a given set of inputs. Therefore, when you are troubleshooting, you may have to apply different sets of inputs in the form of 0s and 1s before you can locate a faulty component or completely test a unit.

Basically, you'll encounter only five kinds of failure when working with the various logic families (MOS, standard TTL, Schottky TTL, etc.):

1. Defective internal logic.
2. Input/output shorted to ground.
3. Input/output shorted to V_{cc}.
4. Input/output open.
5. A short between the IC pins.

The ways in which a failure occurs in the various logic families listed are similar, although some logic families may be more prone to one particular type of failure. For example, a certain IC may be more likely to fail due to defective internal logic because of poor manufacturing practice. Or, you may have more trouble with shorting between pins, due to poor troubleshooting techniques. However, whatever the cause, the trouble symptoms can be identified as follows:

The first failure, that of defective internal logic, will result in erratic readings when you measure the circuit parameters. For example, you may read 0 logic on the output when it's supposed to be a logic 1, or vice versa. Of course, an erratic output such as this will probably bring all operations to a halt if the IC is operating in some type of equipment. To see why you will experience erratic readings, let's examine the failure of components within the input circuitry of a single gate of a 2-input TTL NAND gate such as shown in Figure 2-81.

Let's say that Q_1, the input transistor, opens. Transistor Q_2 will then start conducting and immediately drive the bottom totem-pole transistor, Q_4, into conduction. This will cause the output to go to a logical 0.

Next, let's assume Q_2 develops an open. In this case, it will cause transistor Q_3 to go into conduction, with the end result a steady state 1 on the

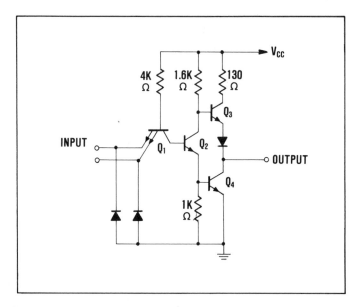

Figure 2-81: Input circuitry for a single gate of a 2-input TTL NAND gate IC.

output. What this comes down to is that you never know what an internal failure will do. It all depends on what happens inside the IC.

The second and third failures, those of input/output shorted to ground or V_{CC}, are also tricky because, in some cases, normal circuit operation continues with only occasional circuit malfunctions. In most cases, however, some expected system operation will come to a halt and you know you have a trouble in that area, which makes it fairly easy to find. In general, with a short to ground all the affected circuits will remain in a logic 0 state. If the short occurs in a V_{CC} circuit, all affected stages will remain permanently in a logic 1 state.

The fourth failure, open input/output, will affect different circuits, depending upon whether it occurs at an input or an output. For example, an open input will probably cause the input to float at some unpredictable level and generally will have no effect on the input signal. On the other hand, an open output will usually cause the TTL device to go to a high logic level somewhere near 1.5 volts for both TTL and DTL. Although this is slightly less than what is correct for a logic 1, you will probably find that all following inputs will act as if they are at a high level. A rule to remember is: In TTL, an open (or floating) input is interpreted by the IC as a logic 1 level.

The fifth failure, a short between two pins, can cause you to want to quit the business. The problem is that when two pins of an IC are shorted together, it causes the outputs of the ICs (which drive those two pins) to attempt to pull the other IC pins to logic 1 or 0. As with any short, the circuit will draw excessive current. In this case, it results in a logic 0 level on both outputs of the driver IC stages. Now, it's possible that when both IC outputs are working together (both high or low level), your test will show the circuit operating properly. But when they are driven into alternate states (one output 0 and the other 1), you will not measure anything but a low state on the output. There is a good chance that the circuit, due to the burned-out components, will finally come to a complete stop. Assuming you spot the trouble before excessive heat destroys some components, the best way to troubleshoot the problem is to have one driving output at logic 1 and the other at logic 0, then check the logic level of the input pins to the driven IC.

Digital Circuit Testing

Let's look at a simple example circuit and examine its operations. The circuit is shown in Figure 2-82 and is the master-slave J-K flip-flop that we have discussed before. This flip-flop is constructed using two 7400 quad 2-input NAND gates and one 7407 hex inverter buffer/driver.

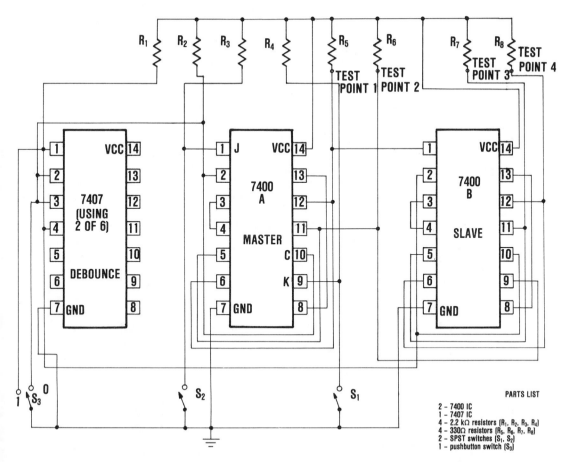

Figure 2-82: Master-slave J-K flip-flop, using two 7400s and a 7407 hex inverter.

If you would like to construct the circuit shown in Figure 2-82 and run some tests, you'll find the ICs are all very inexpensive. Switches S₁ and S₂ (on the J and K inputs) are single-pole-single-throw (SPST) toggle switches, readily available in almost any electronics supply store.

Switch 3 should be a snap-acting push-button switch. This switch will produce a clock waveform when you press the switch button and is *debounced* by the 7407 IC and its associated circuit. Using a manual switch trigger pulse will most likely result in oscillations and indeterminate triggering of the master section IC, if you do not include the clock debouncing circuit. The *master section* is made up using the 7400 IC labeled (A). The *slave section* is made up using the 7400 IC labeled (B).

After the wiring is complete *(leave the power off until the wiring is completed)*, you can proceed with the testing. You'll need a voltmeter and clip leads to connect to the test points shown in Figure 2-82 or, if you have them available, you can connect LEDs in series with resistors R_5, R_6, R_7, and R_8 (break the line at the test points to insert the LEDs).

The next thing you'll need to perform the tests is an appropriate truth table. The one shown in Table 2-31 can be used to set up the input switches *before the clock pulse is initiated*.

S_1	S_2	S_3	Tp_3	Tp_4	MODE
0	0	P	Q_{t-1}	$\overline{Q_{t-1}}$	MEMORY
1	0	P	1	0	S_1 SET
0	1	P	0	1	S_2 RESET
1	1	P	1	1	INVALID

P = S₃ PRESSED FOR COMPLETE CLOCK PULSE WAVEFORM

Table 2-31: Truth table to be used for testing the master-slave flip-flop shown in Figure 2-82.

Procedure:

Step 1. First, do not touch the clock switch. Set switch S₁ to ON (1), and switch S₂ to OFF (0). You should not see any output at the slave section. The master section is in its memory mode at this time.

Step 2. Now, depress the clock switch button (S₃). You should see an instantaneous reading of logic level 1 on test point 2 and logic level 0 on test point 1.

Step 3. Without releasing the switch, check test points 3 and 4. There should be no change in state. You have just placed the slave section in its memory mode.

Step 4. Next, release the clock switch button. This action should cause test point 3 to go to logic level 1 and test point 4 to go to logic level 0. This step gated the slave section which, in turn, produced these outputs and completed the cycle, i.e., stepped the data through the master-slave sections.

In summary, what you should have learned from this testing experiment is: The master section of the master-slave flip-flop responds to the J-K inputs only while the clock input is at logic 1. The slave section will only respond when the clock input goes back to logic 0. We could have used an inexpensive master-slave IC such as the 7473 dual J-K for this test. However, you would not have had access to the outputs of the master sections. Using the circuit we did, allowed you to see each action step-by-step.

Using Your Oscilloscope to Test Digital Circuits

An oscilloscope can find much use in digital testing, especially if you have a dual-trace scope with triggered sweep. However, you'll generally find the older scopes unsatisfactory because neither time nor frequency measurements can be made and, in most cases, the vertical axis is not calibrated, preventing you from making threshold measurements.

When using a scope to check digital pulses, it is important to realize that just because you see a good-looking digital signal on the viewing screen, it does not necessarily mean the circuit is performing properly. To understand why this statement is true, let's review the threshold levels of the TTL logic family. The low threshold is 0.4 volts and the high threshold is 2.4 volts. If you measure a pulse *amplitude* of *less* than 0.4 volts, it is a valid logic 0 level. Or, if you measure a pulse *amplitude* of *more* than 2.4 volts, it is a valid logic 1 level. Now, we can again say that when you are viewing logic levels, absolute amplitudes are unimportant. In fact, when using an oscilloscope, you must check to see if the data is a *valid logic level,* as has been explained.

The scope is excellent for checking pulses for distortion. When you are viewing a pulse or pulse train, look for preshoot, overshoot, delay in rise or fall time, too long or short a pulse width, and a delay between input pulses to a logic circuit, any one of which could cause a trouble in the circuit you are troubleshooting. A dual-trace scope is especially useful when making all these checks. Incidentally, there are oscilloscope accessories you can connect to the input of your scope that will enable you to check several input and output pulse trains at the same time.

Regardless of low-priced scope limitations, you can use them when working with digital equipment, to check some circuits. For example, if the signal you are checking is of a moderately high frequency, a low-cost (narrow bandwidth) scope may perform quite well. When you are attempting to observe a pulsed signal in logic circuits, you'll get better results with a x-10 (low capacitance) probe. Keep the probe ground clip as close to the point of measurement as you can and, by all means, keep the ground lead as short as possible. Another hint: *Do not* use coax for direct connections. Always use the probe. If this seems like strange advice, there is a reason. Coax cable can introduce excessive capacitance into your test setup (the longer the coax lead, the more capacitance), which, in turn, can temporarily (during testing time) disable circuits such as flip-flops, etc. By the way, to play it safe, every time you move the x-10 probe from one test to another, readjust the probe. With a compensated probe, you can encounter considerable error in your readings (when working with high frequencies) if you don't adjust it properly. Figure 2-83 shows several important pulse characteristics that you can examine with an oscilloscope that has both good *vertical deflection sensitivity* and *vertical bandwidth.*

The pulse parameters shown in Figure 2-83 need some clarification. To begin with, all leading and trailing edges of pulses slope. It is not possible for them to rise or fall in zero time, even though they seem to do so when displayed on an oscilloscope set to a slow sweep.

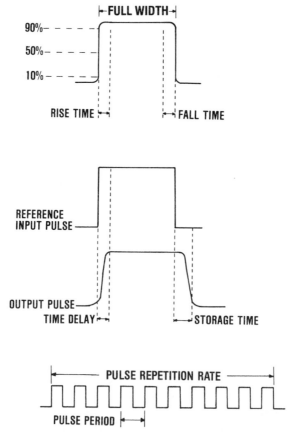

Figure 2-83: Important digital pulse characteristics that can be examined with a good quality oscilloscope.

Risetime is the interval of time required for the leading edge of a pulse to rise from 10 percent to 90 percent of the pulse amplitude. *Falltime* is the interval of time required for the trailing edge of a pulse to decrease from 90 percent to 10 percent of the *pulse* amplitude. The pulse amplitude is the voltage level of the *middle* of the pulse, and approximates to its average level. The peak amplitude is the voltage level of the highest point of the pulse. This is usually the tip of the leading edge, which often overshoots.

The *pulse width* (or duration) is measured horizontally between the 50 percent points of the leading and trailing edges. The *pulse storage time* is the time interval from the 90 percent point on the trailing edge of the *input* pulse to the 90 percent point on the trailing edge of the *output* pulse. Both pulses should be viewed together on the screen of a dual-trace oscilloscope, adjusted so that they have the same amplitude, and the sweep is triggered externally by the input pulse.

The *time delay* is the time interval between the 90 percent point on the leading edge of the input pulse and the 90 percent point on the leading edge of the output pulse, viewed under the same conditions.

Getting the Most Out of a Current Tracer

Your best solution to a wired-OR/wired-AND trouble (as well as shorts and opens when you are working on PC cards) is to use a current tracer. The current tracer operates on the principle that whatever is driving a shorted (low impedance) point must be delivering the greatest amount of the current. Therefore, you should be able to trace the path of the current directly to the low impedance point. A good rule to follow is: If the troubleshooting symptoms indicate an excessive current flow, use of a current tracer is indicated. Figure 2-84 shows the wired-OR/wired-AND configuration (the connection of two or more open collector or tri-state logic outputs to a common bus, so that any 1 can pull the bus down to 0 level).

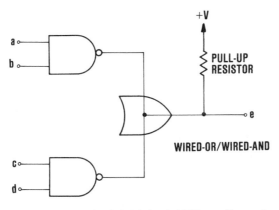

Figure 2-84: Wired-OR/wired-AND configuration.

To detect a wired-OR/wired-AND fault, you can align the mark on a current tracer tip along the length of a printed circuit trace, near the pull-up resistor, and adjust the sensitivity control until the indicator just lights. Then you can move the current tracer along the trace (line), or place it directly on the gate's output pins. Only the malfunctioning gate will cause the indicator to light up.

As you can see from this, although a current tracer looks somewhat like a logic probe and logic pulser, having power leads on one end and a tip on the other, it operates quite differently. The tip does not have to make direct contact with the circuit under test during use. The tip contains a magnetic sensor that is used to monitor the field produced by the current flow in the circuit you are checking. Of course, this makes the instrument excellent for finding low impedance faults. Also, if you are not getting an input signal, you can place a logic pulser at the driving point (pin, etc.). Some of the faults you can locate with a pulser and current tracer are:

1. Shorted inputs of ICs.
2. Bad solder jobs on PC boards (shorted lines due to solder bridges).
3. Shorts between conductors in connections, cables, test leads, and other similar transmission lines.

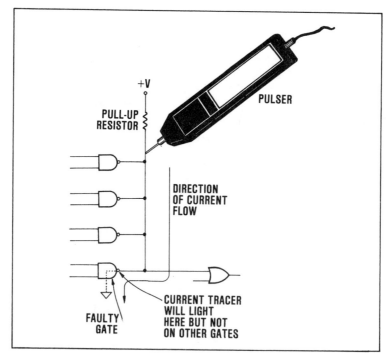

Figure 2-85: Test setup for locating wired-AND troubles.

4. V_{CC}-to-ground shorts.
5. Opens and shorts in a PC card.
6. Signal lines between ICs.
7. When a buffer is driving numerous inputs and one input is shorted to ground.
8. Stuck wired-AND gates.

The last item, number 8, can be very difficult to find without a current tracer. Typically, one of the open collector gates connected in the wired-AND configuration may still continue to draw current after it is supposedly non-conducting. Referring to Figure 2-85, you'll see basically what is happening during the trouble and how to locate wired-AND problems with a current tracer/logic pulser combination.

SECTION 2.15: LIST OF CRITICAL TTL AND CMOS USE PROCEDURES

Even though TTL chips are relatively easy to use and the following tricks and procedures have been explained throughout this book it is extremely important that you review them from time-to-time (particularly if you are new to the field of digital electronics).

1. When working with inputs, do not let the input voltage exceed the supply voltage (generally, $+5.0V$) and do not permit the input to go below ground potential (less than 0 volts).

2. When selecting a TTL circuit power supply, always choose a *regulated supply*. In almost every case, standard TTL ICs are guaranteed by the manufacturer only if V_{CC} is operated between $+4.75$ and $+5.25$ volts, i.e., 5 percent of a nominal 5V. CMOS voltage range can be from $+3$ to $+15$ volts.

3. As a rule of thumb, use number 20 AWG wire for all TTL power connections. Also, use a large electrolytic bypass capacitor (up to 100 μF) between the IC's supply input (V_{CC}) and ground. Keep your capacitor leads short and mount it from the IC's V_{CC} pin to ground, if possible. However, if more than one IC is connected to a common bus line, a single capacitor can be used. Connect the capacitor between the common $+5V$ line and ground. The purpose of this bypass capacitor is to absorb the large current spikes created by the switching action of the totem-pole output circuit.

4. Keep all wires that carry signals as short as practical. Also, do not bundle (make a wiring harness) these wires. The problem you are trying to avoid is magnetic coupling (cross-talk).

5. Use a debounce circuit when setting up digital systems that require mechanical switches.

6. Connect all unused standard TTL *inputs* to the positive supply voltage through a 1k resistor. Or tie them to other similar inputs. It is

extremely important for you to realize that *every unused input pin* of a CMOS IC must be connected to either V_{DD} (the positive supply voltage) or V_{ss} (the negative supply voltage—generally ground).

7. Use a pull-up resistor for a larger high level output. Typically, a TTL gate's high level output pulse is 3.3 volts. You can increase this to +5 volts by adding a 2.2k resistor between a +5 volt source and the gate's output lead.

8. Use an output transistor if you need more current to drive a following unit. Place a 1k resistor between the base of the output transistor and the gate's output lead.

9. CMOS handling precautions:

   a. Do not store CMOS chips in plastic trays, bags or foam.
   b. Always short the pins of a MOS chip. Wrap fine wire or, easier, wrap aluminum foil around the chip so that all pins are shorted together.
   c. Do not touch the pins of CMOS ICs.
   d. When working with CMOS chips, use IC sockets. Or, use a grounded soldering iron.
   e. Check pin configurations of CMOS chips before substituting them for a TTL IC. Some are pin-for-pin compatible, some are not! Example, you can not directly replace a CMOS 4011 (quad 2-input NAND gate) for a TTL 7400 (quad 2-input NAND gate) without rewiring the IC socket.

10. Be sure to check IC specifications. For example, the output of the 7400 (a standard TTL gate) will sink 16 mA in the low state and source 800 μA in the high state. The 74LS00 (a Schottky TTL) will sink 8 mA in the low state and source 400 μA in the high.

SECTION 3

Practical Guide to Servicing
Solid State Circuits and Devices

SECTION 3.1: SERVICING LINEAR/ANALOG
INTEGRATED CIRCUITS

It has been stated that a *linear* amplifier is an amplifier that faithfully reproduces the input waveform on its output. As an example of a linear IC circuit you might encounter, we will use Motorola's general purpose transistor array CA3054. The pin connections and transistor diagram are shown in Figure 3-1. The CA3054 has six NPN transistors that comprise the amplifiers and are useful from dc to 120 megahertz (MHz).

The CA3054 consists of a single slice of silicon substrata on which the integrated circuit is built; hence, it is called a *monolithic integrated circuit*. The monolithic construction of this IC provides close electrical and thermal matching of the amplifiers, which makes it particularly useful in dual-channel applications where performance of two channels is required. Referring to Figure 3-1, you will see that the CA3054 is a *dual independent differential amplifier*. Pins 2 and 13 are the two inputs to the differential amplifier on the bottom half of the IC.

In case you are not familiar with a differential amplifier, it is a circuit that amplifies the *difference* between two input signals, but effectively suppresses all *like* voltages or currents on these inputs. When properly connected, a differential amplifier (such as either of these) will produce an output only when there is a difference in signals at the inputs.

You'll probably find differential amplifiers in oscilloscopes, electronic meters, recording instrument amplifiers, and for data transmission in electrically noisy environments because signals common to both inputs (for example, pins 2 and 13 in Figure 3-1) are eliminated or greatly reduced. Such inputs are known as a *common mode signal*.

Frequently, a differential amplifier is used as an input circuit of an operational amplifier (for instance, Motorola's LM124 series). The ability of the differential amplifier to prevent common mode signals from passing on to the next amplifier (the OP AMP), is one of its most useful characteristics.

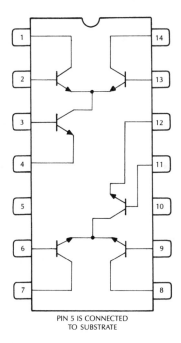

PIN 5 IS CONNECTED
TO SUBSTRATE

Figure 3-1: Pin connections for Motorola's linear integrated general purpose transistor array CA3054.

Of course, as we have said, it is a linear amplifier; therefore, all desirable signals are amplified replicas of the difference signal (the desired signal), and these are fed to the next stage.

Other integrated circuits that are typically called linear are: voltage regulators, operational amplifiers, consumer circuits such as television circuits, TV games/displays, and certain switches, phase-locked loops, etc. However, it should be pointed out that strictly speaking, some of these ICs (such as operational amplifiers) are really performing analog functions rather than linear. Let's now review the basic concepts of voltage regulation.

Guidelines for Servicing Today's Voltage Regulators

One of the characteristics of all voltage regulators is known as *load voltage regulation*. This characteristic refers to the regulator's ability to maintain a nearly constant voltage supplied to the load under all conditions (variable load conditions).

Assuming that your shop temperature is not too hot or cold—and stays constant—and that your workbench ac line voltage remains at the same voltage level during the measurement, the load voltage regulation is expressed as a percentage and calculated (after you have made both measurements required) by using the formula

$$\text{load voltage regulation} = \frac{\text{no load voltage measurement - full load voltage measurement}}{\text{no load voltage measurement}} \times 100$$

If you try several different voltage values for no-load and full-load, you will quickly find out that the lower the load voltage regulation values are, the better the dc power supply stabilization is. For example, suppose that you have a voltage-regulated supply that is providing a full-load current of 7 amperes to a power amplifier circuit. With the amplifier circuit disconnected (no load), the dc power supply output voltage is 38 volts (no load voltage). With the 7-ampere amplifier load connected to the power supply output terminals, the voltage drops to 37.9 volts (full-load voltage). The load voltage regulation of your power supply is

$$[(38.0 - 37.9) / 38.0] \times 100 = \text{approx. } 0.26\%$$

Although this is just an example, you should be aware of the heat created at current levels such as 7 amperes. In general, you are fairly safe using the standard heat sink methods (up to current levels of about 5 amperes). But, at current levels such as 7 amperes and above, it becomes increasingly necessary to use a blower or fan in addition to the heat sinks.

If you measure the load-voltage regulation under different load conditions (more like real world conditions), you will find that the load-voltage regulation value for a basic dc supply will not remain constant. For instance, if you measure the voltage output at 25, 50, 75, and 100 percent of full load, you will find that the relative output (full load to no load) will not change in a linear manner. In fact, as you increase the load, the change in voltage readings will increase. The full-load voltage will decrease as the load is

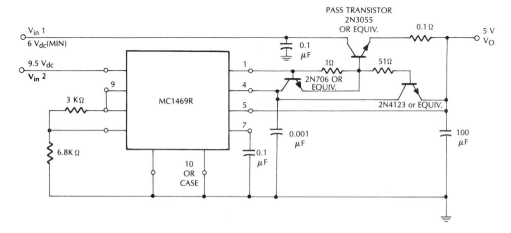

Figure 3-2: Five-volt, 5-ampere regulator based on Motorola's positive voltage regulator IC.

increased. Or, to put it another way, your calculated load voltage regulation will show the power supply to have a higher (worse stabilization) load-voltage regulation.

Because of the wide use of computers, the need for high current regulator circuits has increased considerably in the past few years. Figure 3-2 shows a 5-volt dc, 5-ampere regulator, using a MC1469 voltage regulator suggested by Motorola.

The circuit shown in Figure 3-2 includes two inputs voltages (V_{in1} and V_{in2}). The V_{in2} supply is an auxiliary power source used to power the IC regulator, and the heavy load current is obtained from this second low voltage power supply. This circuit can supply current up to 5.0 ampere to loads such as computers, etc., using an input voltage at V_{in1} (pass transistor collector voltage) of 6.0 volts minimum. Furthermore, the pass transistor is limited to 5.0 ampere by the added short circuit current network in its emitter.

To go one step farther, Figure 3-3 is a schematic of how two IC regulators can be wired to form a ± 15 V, ±400 mA complementary voltage regulator. For systems such as shown, you must use a positive and a negative regulator. For example, Motorola's MC1469R (positive) and MC1463R (negative) are two that are available.

The wiring diagram shown in Figure 3-3 is actually what is called a

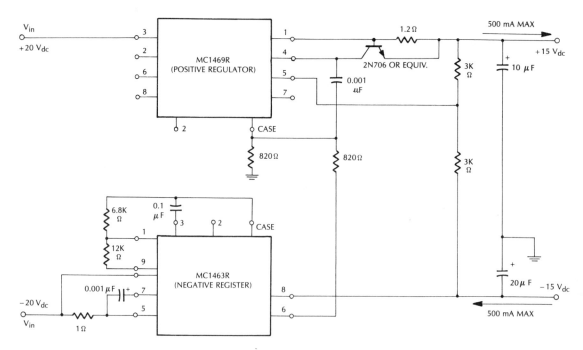

Figure 3-3: Complementary voltage regulator using Motorola's positive and negative regulator IC's, MC1469R and MC1463R.

complementary rocking voltage regulator. This means that both supplies arrive at the same voltage level simultaneously. As an example, when $+V_0$ equals zero output, $-V_0$ must equal zero output. The MC1469R (the positive regulator) is placed in the tracking mode by grounding pin 6 (one side of an internal differential amplifier) and connecting pin 5 (the other side of an internal differential amplifier) at the junction of the two k ohm resistors (see Figure 3-3). The differential amplifier controls the *series pass transistor* (2N706), assuring that the voltage at pin 5 will always be zero. This, in turn, means that you should measure $+V_0 = -V_0$ when pin 5 equals zero.

Switchmode Regulators

The variable resistor action of a series pass transistor is widely used today to regulate the output voltage of a dc supply. But another popular way to control or stabilize output current or voltage is to use a *switching device*, in place of the series pass element, to vary dc output level. Because of certain advantages, switchmode power supplies have become very popular. Some of the advantages are:

1. The series pass type regulator must have the input voltage higher than the output but, by the use of simple jumper wires, the output voltage in a switching regulator can be stepped up, stepped down, or inverted to the opposite polarity.
2. Perhaps the most important advantage: all switching regulators have high efficiency for all input and output conditions (up to 90 percent), and are excellent to use where size is a factor.

Nevertheless, there are disadvantages to go along with the advantages. For example, the following list includes some of the main characteristics of switching type regulators and contrasts their performance with that of the linear units:

1. In general, switchmode regulators' response time to rapid changes in load current is poor compared with the series pass units we have discussed.
2. Unless you use input filters, switchmode regulators can generate electrical noise and electro-magnetic radiation.
3. The advantages of the switchmode regulators are
   a. They can be built small and lightweight.
   b. Often, thermal considerations are of little importance.
   c. They can serve the unique function of a dc-to-dc transformer.
   d. They can be driven with poorly filtered dc.

This all translates into power supplies with small transformers, minimum cooling, and inexpensive operating cost (low power consumption and high efficiency).

Servicing Basic Switchmode Configurations

Although switchmode regulators are listed as linear ICs in manufacturers' data books, switchmode regulators are based on using transistors in a nonlinear fashion. The unregulated input voltage is "chopped" with a saturated transistor and this energy is stored in an inductor and capacitor. This stored energy is then supplied to the load as needed. Figure 3-4 shows a switchmode step-down regulator (more properly called a *pulse-width modulated step-down converter*).

Regulators of this type (switching) provide switching after rectification, and therefore are sometimes called *dc switch regulators*. In general, the dc load current is interrupted at a fixed frequency rate determined by the value of an external resistor, R_T, (pin 6), and an external capacitor, C_T, (pin 5). See Figure 3-4.

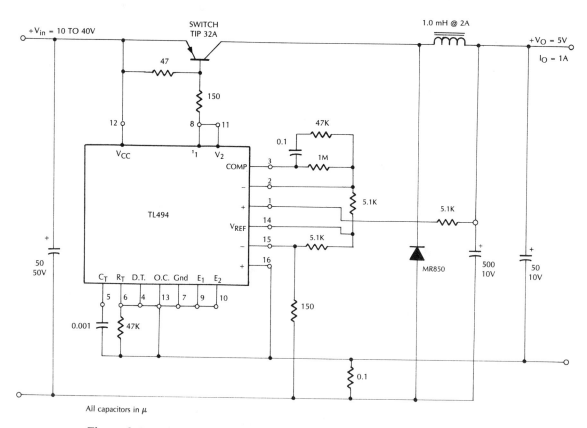

Figure 3-4: Pulse-width modulated step-down converter.

The internal linear oscillator frequency is determined by

$$f_{osc} = 1.1/(R_T C_T).$$

Using the values given for R_T and C_T in Figure 3-4, the frequency of the internal sawtooth oscillator works out to be approximately 23.4 kHz.

In the circuit shown, the control lines, pin 4 dead time control (D.T.), pin 13 output control (O.C.), pins 9 and 10 (emitters of two internal output transistors used in the process of pulse-width modulation), are all tied to ground (pin 7).

Now, let's examine switching regulators such as this one (and other positive regulators) in the most general case. As you have seen, the dc load current is chopped at an audio frequency rate and we can now say that it is controlled by variation of duty cycle. A subsequent inductor is required (see Figure 3-4, 1.0 mH @ 2A inductor) to average the dc level. It also helps remove harmonics. The basic diagram in Figure 3-5 illustrates the fundamental concept of dc switching regulators.

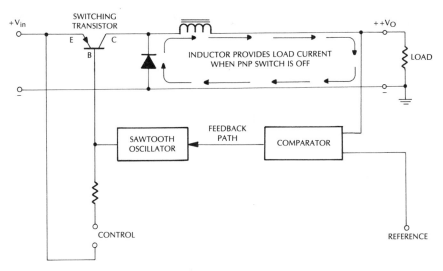

Figure 3-5: Basic functional diagram of a step-down dc switching regulator.

Figure 3-6 shows two other possible arrangements of a switching regulator: a step-up, where the output voltage is greater than the input voltage, and an inverting type, which will produce a negative output for a positive input.

If one of the points is more positive than the other, it is easy to see (Figure 3-6) that load current must be flowing from the left-hand side of the diagram ($+V_{in}$) to the right-hand side ($++V_o$). On the other hand, the electron flow in the inverter is in the opposite direction, from $-V_o$ to $+V_{in}$, and down through the inductor because, in this case, ground is more positive than the regulator output ($-V_o$).

Currently, there are various monolithic switchmode regulator ICs on the market. However, these three—step-down, step-up, and inverting types—are the basic regulators that you will most often come into contact with.

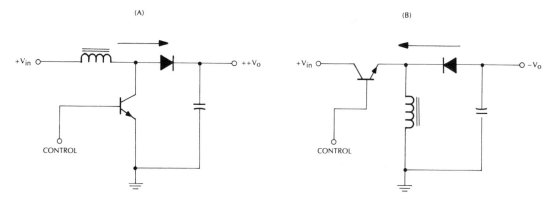

Figure 3-6: Basic diagram of a step-up and inverting type switchmode regulator. (A) step-up, (B) inverting.

Operational Amplifier Servicing Guide

Today, the availability of inexpensive operational amplifiers has made them a practical replacement for *any low-frequency amplifier.* By no means are these OP AMPs restricted to linear amplifiers. In fact, you'll find they are used in an incredible variety of dc and audio uses. Just a few examples are active filter circuits, precision rectifiers, integrators, ramp generators, electronic music circuits, and many more.

By now, everyone is probably more or less familiar with the 741 and most of its improved offspring such as the 5558 (Signetics), MC1458 (Motorola), 4136 (Raytheon), and the improved LMM318 (Advanced Micro Devices, and National). In general, you'll find that most of these OP AMPs are available at reasonable cost from electronics parts suppliers. They also are the type of OP AMPs you may encounter during troubleshooting electronics equipment manufactured in the past few years.

There are OP AMP ICs that incorporate highly matched junction field-effect transistor (JFET) devices on the same chip with standard bipolar transistors. The JFET devices enhance the input characteristics of these OP AMPs (known as monolithic JFET operational amplifiers) considerably more than a conventional amplifier.

If more gain is needed than can be obtained from a single differential amplifier, the manufacturer cascades several more stages to build an OP AMP. This provides both common-mode rejection and high gain. A schematic symbol of a conventional OP AMP and a simplified block diagram are shown in Figure 3-7.

By referring to Figure 3-7, you'll notice there are two inputs to the differential input stage of the OP AMP. When reading a schematic, you'll normally find one input marked with a minus sign. This input is called the *inverting input.* The other input is usually marked with a plus sign and is called the *non-inverting input.* The schematic symbols of (+) and (−) on the

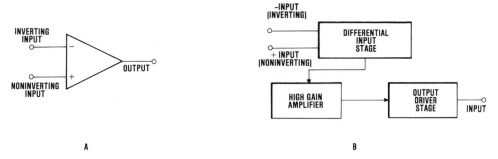

Figure 3-7: OP AMP schematic symbol (A), and simplified block diagram of the interior of an OP AMP IC (B).

signal input circuit have nothing to do with the power supply input. You may find the voltage supply pins marked V + and V −, while other manufacturers may use V_{EE} and V_{CC} to represent negative and positive respectively. In other words, you must look at the schematic carefully before applying the amplifier power source.

In the so-called good old days, we usually thought that any power supply terminal marked negative was an earth ground point. This is still true when working with many circuits, but not for all OP AMPs. There is no ground connection for some OP AMPs and you'll find many use split supplies, for example, one + 15 V line and one − 15 V line. Furthermore, the input signal level must be restricted to working with significantly less than the full amplitude of either of the two supply voltages. To illustrate, in one case (a 741 OP AMP), the two supply voltages are + 15 V and − 15 V. The input signal level is limited to ± 12 volts. You'll find this specification called the *common mode range.*

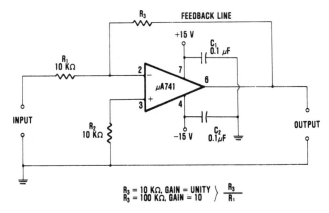

Figure 3-8: A typical example of a feedback OP AMP operating in the linear (inverting) mode. The gain of such a circuit is determined by the ratio of external feedback resistance. Circuits may differ from the one shown by having more components, but the basic operation is still the same.

If the circuit you're working on is using both inputs, it is working in the *differential mode*. But if you find that one of the inputs is grounded (or tied to some other reference), it's said to be working in the single-ended mode. Next, when you are checking the schematic of a piece of gear using OP AMPs, look at the feedback lines. If you find that the feedback is from the output to the positive input, it's probably a digital operation. On the other hand, if the feedback is from the output to the negative input, the circuit is set up for a linear operation. The gain of the circuit depends on the ratio of R_3 and R_1, as indicated (see Figure 3-8).

Glossary of OP AMP Terms Needed for Servicing

Input Offset Voltage——This is the slight voltage difference (a millivolt or so) between the OP AMP inputs.

Input Offset Current——This refers to the difference in the two currents (measured at the two input terminals).

Input Voltage Range——Watch this parameter! It is the range of voltage you can safely apply to the input without damaging the device.

Output Voltage Swing——This is the peak output voltage range, referred to zero, that you can obtain without voltage clipping.

Current Offset——At very low impedance levels (1 k ohm or so), you can usually ignore the input currents and input offset current. If the impedance is higher, current may become important. When you are working with low impedance values, don't worry about anything except input offset voltage. However, when the impedance is greater, you will have to take both voltage and current into consideration. Current problems are usually caused by a mismatch in the input state.

Slew Rate——This is the maximum rate of change of the output voltage with respect to time that the OP AMP is capable of producing and still operate in the linear mode. You'll usually find it listed in volts per microsecond.

It must be remembered that bandwidth, slew rate, output voltage swing, output current, and output power of an OP AMP are all interrelated. Furthermore, these characteristics are frequency- and temperature-dependent. Because of all these problems, you'll find circuits for reducing offset current, frequency compensation, and bias current are included internally and externally with every OP AMP.

How to Make Voltage Gain Measurements

Test Equipment:
   See **Comments.**

Test Setup:
   See **Comments.**

Comments:

Basically, a voltage gain measurement for an OP AMP is the same as for an audio amplifier. Just remember, don't overdrive any IC OP AMP. It's possible to damage the IC and, almost without question, you will end up with erroneous readings.

Procedure:

Step 1. Apply the manufacturer's maximum rated input, then measure the output. If possible, check the output with a scope.

Step 2. If you see clipping taking place when you apply the recommended input signal (before you make a measurement), reduce the input signal until the clipping stops.

Step 3. Next, increase the input signal frequency until the voltage gain drops 3 dB from the initial low frequency value. This is the OP AMP's *open-loop bandwidth* and voltage gain.

Step 4. To check the *closed-loop gain,* use the same procedure, except now the IC must have a feedback circuit. The closed-loop gain of an OP AMP is dependent upon the ratio of the feedback and input resistance (see Figure 3-8). When you check an OP AMP's closed-loop voltage gain and bandwidth, you will find lower readings than your open-loop measurement values. In other words, the frequency response won't be as good and the gain will be less.

Making OP AMP Input/Output Impedance Measurements

Test Equipment:
See **Comments.**

Test Setup:
See **Comments.**

Comments:

As we explained, when you measure the characteristics of an OP AMP, it must be remembered that the open-loop values will be higher than closed-loop values. This also is true of impedance values, i.e., open-loop impedance will differ from closed-loop impedance. To find the input impedance with the OP AMP operating into a load, construct a test setup like the one shown in Figure 3-9. Incidentally, if you are checking the same OP AMP as the one used in the preceding measurement (voltage gain), be sure that you use the same frequency, or frequencies, load, and test gear for this measurement as you did for that test.

Procedure:

Step 1. After you have the setup shown in Figure 3-9, adjust the sine-wave generator to the frequency at which the OP AMP is to be operated.

Step 2. Now, adjust your variable resistor (shown as R in the drawing) until you read exactly the same signal voltage on each of the voltmeters.

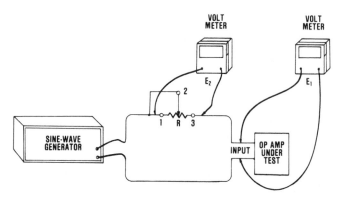

Figure 3-9: Test setup for measuring the input impedance of an OP AMP under load.

Step 3. Finally, disconnect the resistor and use the ohmmeter to measure the resistance between resistor terminals 2 and 3. What you read with your ohmmeter is the dynamic input impedance of the OP AMP under test. *Note:* If the test frequency is very high, be sure to use a non-inductive load resistor. Another point: Obviously, you can use one voltmeter in place of the two and, to speed things up, a resistor substitution box can be used in place of a single variable resistor. But don't forget, the accuracy of the measurement will depend on the tolerance of the resistor used as well as the accuracy of the ohmmeter.

How to Make Noise Measurements

Test Equipment:
    See **Comments.**

Test Setup:
    See **Comments.**

Comments.

Noise measurements for OP AMPs are the same as for audio amplifiers. However, as with any audio amplifier, the frequency spectrum or random noise will show different output levels at different frequencies, and is dependent upon the amplifier under test. By referring to Figure 3-10, you will notice that (A) shows the voltage amplitude constant from 0 Hz to 10 Hz and beyond. This is often called *white noise.* You may see this type of noise referred to as *Johnson noise* but, in the strictest sense, Johnson noise is the noise generated by any resistor at a temperature of absolute zero in degrees Kelvin.

In practice, a perfect noise output level isn't common because an amplifier will affect the output at various frequencies. For example, in Figure 3-10 (B) the amplifier has higher gain at the low frequencies than at the highs. This type of noise spectrum is often called *red noise.*

On the other hand, if the amplifier has better gain in the high frequency region, you will find the output graph would look like (C). This is *blue noise.* Next, the graph in (D) shows what is called *pink noise.*

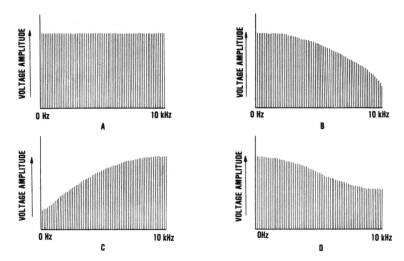

Figure 3-10: Graphs of an audio amplifier's effect on white noise spectra. (A) white noise, (B) red noise, (C) blue noise and (D) pink noise.

Typical noise for a low-cost OP AMP such as the 741, when referred to the input, is 10 μV, which is not particularly good. Thus, if your OP AMP has a gain of 10, you should measure 100 μV on the output. At a gain of 100, you would have 1 millivolt out, and so on. What this means is that you will need a pretty good scope to make a noise measurement. The test setup is shown in Figure 3-11.

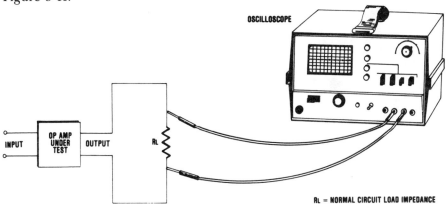

Figure 3-11: Test setup for measuring an OP AMP's noise level. The scope should be capable of a measurable deflection with 1 millivolt or less. Generally, this measurement is made with the OP AMP operating in the open-loop mode.

Procedure:

Step 1. Adjust the gain control on your scope until you have no noise showing. Your scope should be showing a light line across the face.

Step 2. Next, increase the scope gain control until you have a measurable peak-to-peak amplitude of the noise off the face of the scope.

Step 3. It is possible to get an erroneous reading due to the scope leads picking up ac hum. To check this, set your scope sync control to operate at 60 Hz. If you can get the pattern to appear stationary, you are picking up 60 Hz interference.

Step 4. OP AMPs can break into oscillation. If you suspect this is happening, short the input to the OP AMP circuit. If you still see hash (sometimes called *hum*), the OP AMP circuit, more than likely, is oscillating.

How to Measure the Input-Offset Voltage and Current of an OP AMP

DC base bias current *must* be provided for both the + and − input to an operational amplifier. In theory, the input bias current should be the same for both inputs. In reality, they should be as close to equal as possible. However, even with the best of matching, there will be a slight difference between this offset voltage and true input signal. A handy rule to remember is: The output offset voltage will equal the input offset voltage times the in-circuit gain of the OP AMP. The test setup is shown in Figure 3-12.

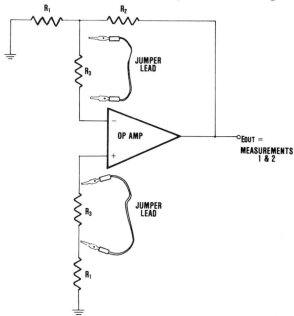

Figure 3-12: Test setup for measuring an OP AMP's input-offset voltage and current.

As a typical example, let's assume that you have a test setup using these resistor values:

$$R_1 = 51 \text{ ohms}, \quad R_2 = 5.1 \text{ k ohms and } R_3 = 100 \text{ k ohms.}$$

Next, under these conditions, we will assume that you measure the two voltage outputs and find them to be $E_1 = 80$ millivolts (jumper leads connected) and $E_2 = 360$ millivolts (jumper leads removed). All you need now are two simple calculations:

input-offset voltage $= E_1/(R_2/R_1)$
$= 80$ mV/(5.1 k ohms/51 ohms)
$= 80$ mV/100 $= 0.8$ mV
input-offset current $= E_2 - E_1/R_3(1+5.1$ k ohm/51 ohm)
$= 280$ mV/100 k $(1 + 100) = 27.7$ nano A

How to Use a Square Wave to Measure the Slew Rate of an OP AMP

The slew rate problem of an OP AMP can be stated simply: The slew rate of an operational amplifier limits large output, high frequency signals. Or, to put it another way, ordinarily, you can't have large output swings and high frequency operation at the same time. Perhaps, the easiest way to observe and measure the slew rate of an OP AMP is to measure the slope of the output wave form of a square-wave input signal, as shown in Figure 3-13. When you make this check, it is important to remember that your square-wave generator *must* have a risetime better than the slew rate capability of the OP AMP under test. For example, the fastest you can normally change the output of a 741 is 0.5 volts per microsecond. Therefore, as long as your square-wave generator has a faster risetime, your measurements will be correct because any distortion you see on the scope will be caused by the OP AMP, not the test equipment.

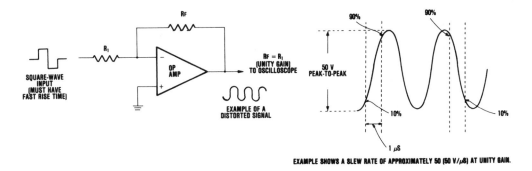

Figure 3-13: Test setup and example slew rate measurement pattern that might be seen on a scope when checking an IC such as an LM318.

Troubleshooting OP AMPs with a Scope

Test Equipment:
See **Comments.**

Test Setup:
See **Comments.**

Comments:
When you hear servicemen speaking of troubleshooting today's solid state equipment, more than likely you will hear at least one of them say, "The oscilloscope is the most valuable tool in the field of electronics measurement." While this may be an overstatement, it's so close to accurate that it is

surprising to find so many technicians using a scope without understanding how to troubleshoot with it effectively. Perhaps one of the reasons for this is that the price of high-performance scopes has decreased dramatically. For about $300.00, the technician can purchase a 10 kHz triggered oscilloscope with a vertical input sensitivity of 10 millivolts per centimeter. Suddenly, we find ourselves face-to-face with items such as x 5 sweep magnifiers, auto and normal trigger level controls, and a host of other specialized functions that, because of cost, did not even exist for most of us just a few years ago. But once you learn to operate the scope, these controls become an asset. On the other hand, if you don't follow some simple rules, troubleshooting OP AMPs with a scope can be troublesome.

A wide variety of useful tests on audio equipment using OP AMPs can be made with a scope. It is a general rule that your test equipment must have performance characteristics equal to, or better than, the OP AMP (or any other device) under test. There are, however, certain exceptions that are possible if you use the right test procedure. For example, for an amplitude nonlinearity check with a scope—using the following procedure—you must first determine the linearity of the scope itself. Large percentages of distortion in any audio waveform are easy to "see," but you really have to look *very* closely to see small percentages of distortion such as illustrated in Figure 3-14.

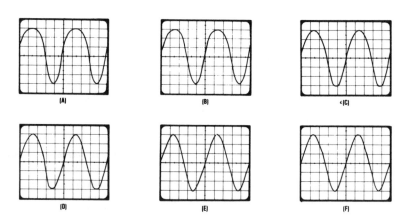

Figure 3-14: These examples clearly show how difficult it is to see various percentages of distortion, using a scope. (A) 20%, (B) 15%, (C) 10%, (D) 5%, (E) 3% and (F) 1%

If you have good eyes, it is possible to estimate the percent of distortion by viewing a sine-wave test pattern on your scope, as shown in Figure 3-14. A good way to improve the effectiveness of a scope measurement is by utilizing Lissajous patterns.

Checking the scope's amplifiers, provides you with a reference pattern for use in evaluating the linearity of the audio amplifier under test. To make this check, connect the output from an audio oscillator to both the vertical and horizontal inputs of your scope, as shown in Figure 3-15.

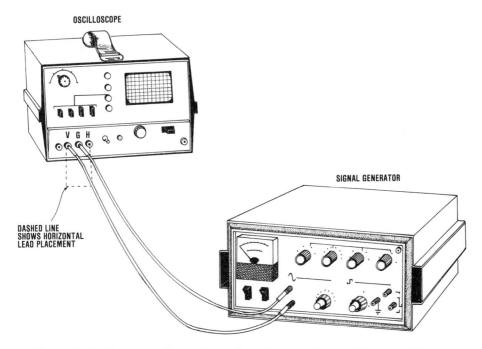

Figure 3-15: Test setup for making a linearity test of an oscilloscope. This test setup provides a reference pattern for testing an OP AMP, etc.

Procedure:

Step 1. Connect the output from your audio signal generator to both vertical and horizontal input terminals of the scope. Don't worry about the linearity of the audio signal source you are using because the waveform presented at its output is of no concern when using this procedure.

Step 2. Next, set the audio signal source operating frequency to approximately 400 Hz.

Step 3. Now you should see a diagonal line displayed on your scope screen. If the scope amplifiers are linear, you should see a perfectly straight line, but if the scope is introducing distortion, the line will be slightly bent, i.e., show some curvature, as in Figure 3-16. The perfectly straight line is what you want for a reference, to make an accurate evaluation of the amplifier under test.

The test setup for using this procedure is shown in Figure 3-17 and the load resistor, R, should be equal to the recommended load impedance for the

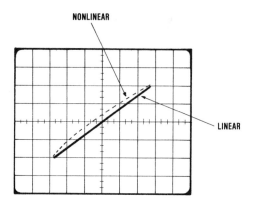

Figure 3-16: Scope presentation of a reference linearity pattern showing a slight curvature. The curved line indicates the scope's amplifiers are nonlinear.

OP AMP. Generally, when testing a low power OP AMP, you don't have to worry about the wattage rating of the load resistor.

Step 4. Finally, observe the pattern on the scope screen. If it is exactly the same as your reference pattern, the OP AMP under test is linear. If you make a low power output measurement and a maximum rated power output measurement, you will usually find that the linearity improves as you reduce power out. *Note:* It may be impossible to see any departure from your reference line if the OP AMP has good performance characteristics.

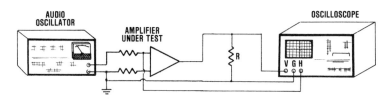

Figure 3-17: Test setup for checking OP AMP linearity, using a scope.

Oscilloscope Measurement Precautions

Most of us tend to assume that an oscilloscope display is telling us what actually is happening in a circuit under test. Believe me (and I am speaking from experience), this may or may not be true! Select the wrong scope probe and it's quite possible that you will see some strange things during certain tests. Take a look at Figure 3-18. You probably can tell, by close examination, that it is a weird-looking square-wave pattern. What's the trouble? It is simply a case of connecting the wrong probe to a scope. What has happened is that a service-type C probe was connected to a high-performance scope. The test signal fed through the service probe to the scope was a 1 MHz square wave with a risetime of 0.02 μsec. What you are seeing is a large amount of overshooting and ringing.

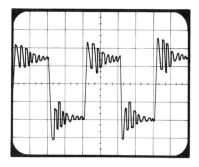

Figure 3-18: An example of overshoot and ringing, caused by using the wrong probe/scope combination.

Unless you take certain precautions, especially at high frequencies, getting the signal to your scope can be the toughest job encountered when making a test. For example, if you fail to match the probe to the scope, it can result in as much as a 50 percent measurement error. Fortunately, there are three rules that will help, under most testing conditions. These are:

1. If loading error is a problem, you can reduce your error to less than 1 percent by selecting a scope/probe combination that has a resistance, looking into the probe, of at least 100 times as great as the signal source impedance.
2. To keep the frequency-related errors down, select a scope/probe combination with a shunt capacitance as small as possible.
3. To reduce phase measurement errors, use low impedance scope/probe combinations.

Always use a low-capacitance probe instead of a direct probe, in order to avoid pattern distortion caused by circuit loading. If you are checking signals in high-frequency circuits, a demodulator probe should be used. But don't make the mistake of using a demodulator probe when a low-capacitance probe should be used. For example, don't use a demodulator probe when checking a TV set in the video amplifier circuits (anywhere after the IF stages).

SECTION 3.2: DIGITAL ICS

In Section 2.1, there are schematics of integrated circuits constructed using transistor-transistor logic (TTL, see Figure 2-1), complementary metal-oxide semiconductor (CMOS, see Figures 2-2 and 2-3), and emitter-coupled logic (see Figure 2-4), plus a brief explanation of Schottky TTL in Section 2.5. Although Schottky TTL uses Schottky diodes to perform the AND function, as explained in Section 2, and emitter-coupled logic is a *non-saturating form of logic* (TTL is a saturation logic, i.e., is either almost totally cut off or on), all essential digital functions are the same regardless of the particular IC family.

As shown in Section 2.1, a gate (or any other device such as a flip-flop, shift register, etc.) operates exactly the same way whether the IC is constructed using ECL or MOS technology. An ECL IC permits very high-speed operation and may have different input and output loading characteristics from its MOS counterpart, but the functions of flip-flops or counters are always the same.

Testing a Digital IC Delay Circuit

Occasions may arise when you need an oscillator with independent control of period and pulse width. The 74122, 74123, (TTL), or the 74C221 (CMOS) dual retriggerable monostable multivibrators with *clear*, and using the circuit shown in Figure 3-19, perform this task very well.

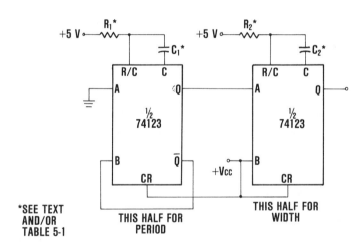

Figure 3-19: This monostable will provide you with complete flexibility in controlling the pulse width, either to lengthen the pulse by retriggering, or to shorten by clearing. Figure 3-20 shows the 74123 IC pin configuration.

The IC has an internal timing resistor that allows you to operate the circuit with only an external capacitor. However, if you use potentiometers for R_1 and R_2, you can construct a low-cost, wide-range pulse generator with lots of control over the output pulse. The output pulse is primarily a function of the external capacitor and resistor you use. For instance, for a capacitor of greater than 1000 pF, the output pulse width (t_w) is defined as:

$$t_w = 0.32 \ R_T \ C_{ext} \ (1 + 0.7/RT)$$

where:

R_T is in k ohms (either internal or external timing resistor)
C_{ext} is in pF
t_w is in ns

For pulse widths when C_{ext} is less than 1000 pF, see Table 3-1.

OUTPUT PULSE WIDTH VS EXTERNAL TIMING CAPACITANCE

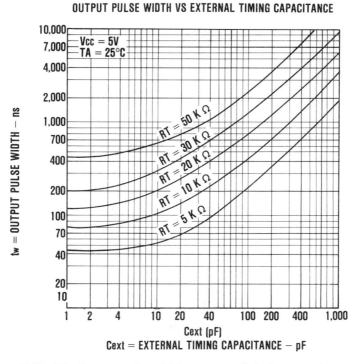

Table 3-1: Output pulse width vs external timing capacitance.

This IC is also retriggerable, as we have said, and like other retriggerable monostable vibrators, it will respond to inputs that arrive while the output is still high from the preceding trigger. This means you can have a train of inputs that will hold the output high until you stop the train of input pulses. When you wire up the circuit shown in Figure 3-19, you will need the 74123 pin configurations. Figure 3-20 shows the IC package with all pins identified.

After you have the IC mounted and wired for operation, your next step is to run tests on the circuit. To do this, you will need a truth table. Inputs and outputs for the 74123 are shown in Table 3-2.

Two other manufacturers' tables that you should have to run this test are

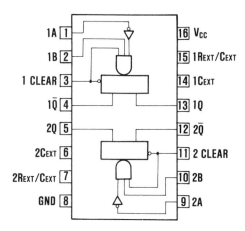

Figure 3-20: Pin configuration for the 74123.

INPUTS		OUTPUTS	
A	B	Q	Q̄
H	X	L	X
X	L	L	H
L	↑	⎍	⎍
↓	H	⎍	⎍

A.   H   = high level (steady state)
      L   = low level (steady state)
      ↑   = transition from low to high level
      ↓   = transition from high to low level
      ⎍   = one high level pulse
      ⎍   = one low-level pulse
      X   = irrelevant (any input, including transitions)

B.   An external timing capacitor may be connected
     between Cext and Rext/Cxt (positive)

Table 3-2: Truth table for a 74123 IC.

the electrical characteristics and switching characteristics. Tables 3-3 and 3-4 include this information.

The greatest single source of problems in this test is false triggering and the next most frequent problem is no triggering at all. IC monostables are very fast and if they can find a way to trigger themselves, believe me, they will. Therefore, when you are testing this circuit, keep your input lines as short as possible. Above all, to prevent stray coupling, keep the input lines as far away as practical from all other lines.

If you have troubles, use a scope with its ground lead connected to your circuit power supply ground. Take a look at your signal ground line to make sure it really is ground. There should not be any detectable voltage or noise (you will see grass across the scope face, rather than a pure straight zero signal line).

ELECTRICAL CHARACTERISTICS

	PARAMETER	TEST CONDITIONS*		MIN	TYP**	MAX	UNIT
V_{IH}	High-level input voltage			2			V
V_{IL}	Low-level input voltage					0.8	V
V_I	Input clamp voltage	$V_{CC} = $ MIN	$I_I = -12$mA			-1.5	V
V_{OH}	High-level output voltage	$V_{CC} = $ MIN See note 1	$I_{OH} = -800\mu$A	2.4			V
V_{OL}	Low-level output voltage	$V_{CC} = $ MIN See note 1	$I_{OL} = 16$mA		0.22	0.4	V
I_I	Input current at maximum input voltage	$V_{CC} = $ MAX	$V_I = 5.5$ V			1	mA
I_{IH}	High-level input current data inputs clear input	$V_{CC} = $ MAX	$V_I = 2.4$ V			40 80	μA
I_{IL}	Low-level input current data inputs clear input	$V_{CC} = $ MAX	$V_I = 0.4$ V			-1.6 -3.2	mA
I_{OS}	Short-circuit output current†	$V_{CC} = $ MAX	See note 1	-10		-40	mA
I_{CC}	Supply current (quiescent or triggered)	$V_{CC} = $ MAX See notes 2,3			46	66	mA

1. Ground Cₑₓₜ to measure Vₒₕ at Q, Vₒₗ at Q, or Iₒₛ at Q. Cₑₓₜ is open to measure Vₒₕ at Q̄, Vₒₗ at Q̄, or Iₒₛ at Q̄.
2. Quiescent Icc is measured (after clearing) with 2.4V applied to all clear and A inputs, B inputs grounded, all outputs open.
3. Icc is measured in the triggered state with 2.4V applied to all clear and B inputs, A inputs grounded, all outputs open. Cₑₓₜ = 0.02μF, and Rₑₓₜ = 25 kΩ. Rᵢₙₜ is open.
* For conditions shown as MIN or MAX, use the value specified under recommended operating conditions.
** All typical values are at Vcc = 5V, Tₐ = 25°C.
† Not more than one output should be stored at a time.

Table 3-3: Electrical characteristics for the 74123 IC.

SWITCHING CHARACTERISTICS

	PARAMETER	TEST CONDITIONS		MIN	TYP	MAX	UNIT
t_{PLH}	Propagation delay time, low-to-high-level Q output, from either A input				22	33	ns
t_{PLH}	Propagation delay time, low-to-high-level Q output, from either B input				19	28	ns
t_{PHL}	Propagation delay time, high-to-low-level Q̄ output, from either A input	$C_{ext} = 0$ $C_L = 15$pF	$R_{ext} = 5$K $R_L = 400$		30	40	ns
t_{PHL}	Propagation delay time, high-to-low-level Q̄ output, from either B input				27	36	ns
t_{PHL}	Propagation delay time, high-to-low-level Q output, from clear input				18	27	ns
t_{PLH}	Propagation delay time, low-to-high-level Q̄ output, from clear input				30	40	ns
t_w(min)	Minimum width of Q output pulse				45	65	ns
t_w	Width of Q output pulse	$C_{ext} = 1000$pF $C_L = 15$pF	$R_{ext} = 10$k $R_L = 400$	3.08	3.42	3.76	μs

Table 3-4: Switching characteristics for the 74123 IC.

Testing a Timer IC

Although the 555 (which we will use as an example) is a timer IC, it can be used as a one-shot, free-running (astable), or gated multivibrator. In fact, this IC can be used in a variety of applications. Figure 3-21 is a 555 timer function diagram that you can use as a guide when working with this, or any similar, IC.

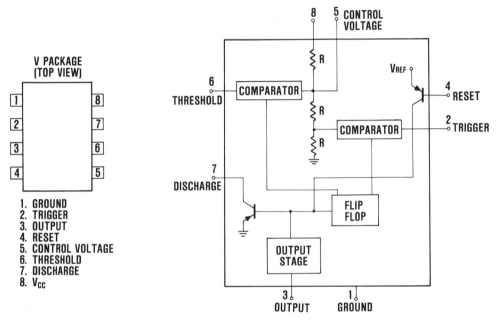

V PACKAGE (TOP VIEW)

1. GROUND
2. TRIGGER
3. OUTPUT
4. RESET
5. CONTROL VOLTAGE
6. THRESHOLD
7. DISCHARGE
8. V_{CC}

Figure 3-21: 555 timer block diagram and V package, top view.

For those who like mathematics, there are four defining equations for the free-running mode when using this IC:

$$\text{duty cycle} = t_2/t_1 = (R_1 + R_2) / (R_1 + 2R_2)$$
$$\text{output high time } (t_1) = 0.693 \, (R_1 + R_2)C$$
$$\text{output low time } (t_2) = 0.693 \, R_2C$$
$$\text{total time } (t_1 + t_2) = 0.693 \, (R_1 + 2R_2)C$$

If you want to try a 555 in the free-running mode, Figure 3-22 shows the basic hookup. The frequency of the circuit can be calculated by using the formula

$$t = 1/T = 1.44/(R_1 + 2R_2)C$$

As you can see by referring to the formula for frequency, if trimmer potentiometers are used for both R_1 and R_2, the frequency (and duty cycle) can be trimmed to your exact requirements. By connecting the output of the

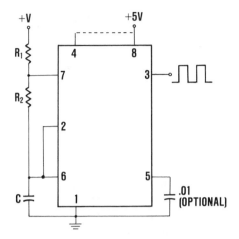

Figure 3-22: Basic connections for using a 555 timer IC as an astable multi-vibrator. See text for R_1, R_2, and C values.

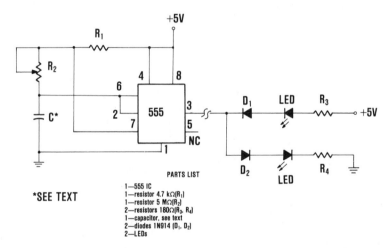

Figure 3-23: Schematic and parts list for testing a 555 timer IC.

basic multivibrator to two LEDs, as shown in Figure 3-23, you can monitor the output.

With any multivibrator, as you know, you will have an on and off output waveform. However, if it is operating at a frequency over 20 HZ, your persistence of vision will make it appear that the two LEDs shown in Figure 3-23 are continuously on. Therefore, in order to see the switching action, you will have to set the operation of the circuit at a very low frequency. This (increasing or decreasing frequency) is merely a matter of changing the value of the capacitor labeled C. You can start off with a capacitor value of 0.022 μF, which should permit you to see a definite on-off action.

Testing the Monostable Operation of the 555 IC

Test Equipment:
> See **Comments.**

Test Setup:
> See **Comments.**

Comments :

In this mode of operation, you will see how the 555 acts when it is wired as a one-shot multivibrator. The wiring of the IC is simple, as shown in Figure 3-24. The built-in discharge stage at pin 7 of the 555 has an internal transistor that holds the external capacitor, C, initially at discharge (due to a short-circuit action by the transistor).

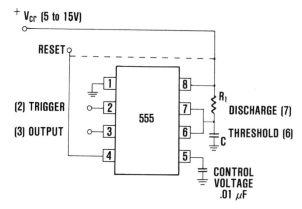

Figure 3-24: Wiring diagram for using a 555 timer IC in the monostable mode of operation. The capacitor connected to pin 5 is essential to reduce noise. Also, when the reset function is not being used, pin 4 should be tied to pin 8 (V_{cc}) to avoid false triggering.

Procedure:

Step 1. To start the operation, apply a negative trigger pulse to pin 2 (trigger pin). The flip-flop circuit (see Figure 3-21) is now set, and this releases the short circuit across the external capacitor C. At this instant, you should find the output (pin 3) changed to a high logic level.

The capacitor now starts to charge. When this charge voltage reaches about 2/3 the voltage you are using for V_{cc} (5 to 15 V), the 555 comparator (see Figure 3-21) resets the flip-flop, which will then discharge the capacitor, C, very quickly and drive the output to a low state. You should be able to view this with a voltmeter, logic probe, or scope. If you are checking with a scope, Figure 3-25 shows the approximate wave patterns you should see at the input,

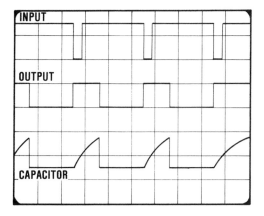

Figure 3-25: Example waveforms that you can expect to see when using an oscilloscope to view a 555 timer wired for monostable operation.

output, and the capacitor voltage. The exact waveforms will depend on what values you use for R and C, the scope sweep time, etc.

Once you trigger this circuit, it will remain in this state until the time you have it set to has elapsed, even if you trigger it again during the set time interval. You will find that the circuit triggers on a negative-going input signal when the level reaches 1/3 of the supply voltage (V_{cc}) you chose to use. However, the time that the output is in the high state depends on what values you chose for R and C. The time that the output is high can be calculated by using the formula t = 1.1 RC. As you can see, 1 megohm and 1 μF will produce a time period of more than a second.

Step 2. Once you have a time period of about a second or so, try applying a negative pulse to both pin 4 and pin 2 (trigger) during the pulse period. You must apply the trigger pulse to both pins simultaneously (reset should not be connected to V_{cc}) when performing this test. This will discharge the external capacitor (C) and start the cycle over again.

Step 3. After completing this part of the test, don't forget to reconnect pin 4 (reset) back to pin 8 (V_{cc}). If you don't, you may have problems with false triggering while trying to operate the circuit as a one-shot.

Cleaning Up Noise or Irregular Signals Using Schmitt-Trigger ICs

You will find that a Schmitt-trigger IC is one of the most reliable and useful of all possible input schemes. One of the reasons is that its overall effect is to clean up noisy or irregular digital signals so that they can be applied to other logic ICs. Furthermore, Schmitt-trigger gates are widely available at comparatively low cost. *Note:* See Section 2.9 for more about Schmitt-trigger ICs.

Figure 3-26 shows the pin configuration and logic symbol for the dual 4-input positive-NAND Schmitt-triggers 7413. Note that the symbol is that of a logic inverter with a hysteresis loop figure drawn in its center. Normally, there is hysteresis between an upper and lower triggering level. The hysteresis, or backlash, which is the difference between the two threshold levels, typically is 800 mV for this IC.

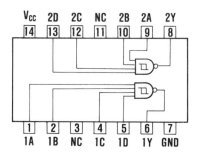

Figure 3-26: Pin configuration and logic symbol for the 7413 4-input positive NAND Schmitt-triggers. These gates were specifically designed for cleaning up noise or irregular digital signals.

As an example of what a Schmitt-trigger can do for you, let's assume that a digital pulse has been sent through a noisy circuit such as a long cable, and arrives at the input looking like the one shown in Figure 3-27. Next, look at the logic levels emerging from the IC—the signal labeled *output*. Notice, it is perfectly clean and ready for application to the next stage. *Note:* A Schmitt-trigger gate cannot be triggered from straight dc levels.

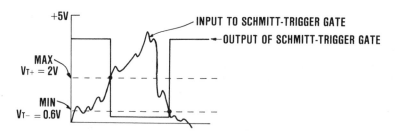

Figure 3-27: Irregular signal input to a Schmitt-trigger gate and its cleaned-up output signal.

There are a number of methods you can use for testing this Schmitt-trigger (i.e., the 7413). There are a variety of IC sockets on the market. The type you use will determine how you mount and/or wire the IC for this test. The most important information you need after the circuit is mounted are the electrical characteristics and the operating conditions for the 7413 IC.

PARAMETER		MIN	NOM	MAX	UNIT
SUPPLY VOLTAGE		4.75	5	5.25	V
FAN-OUT FROM EACH OUTPUT	HIGH LOGIC LEVEL			20	
	LOW LOGIC LEVEL			10	
OPERATING FREE—AIR TEMPERATURE, T$_A$		0	25	70	°C
MAXIMUM INPUT RISE AND FALL TIMES		NO RESTRICTION			

Table 3-5: Recommended operating conditions for a dual NAND Schmitt-trigger (7413 IC).

The recommended operating conditions are shown in Table 3-5. *Leave the power off until the wiring and/or mounting is complete.*

To test the device, we will use a simple scheme that will change common house current (60 Hz) into a digital signal (1 or 0 binary). Figure 3-28 shows a schematic diagram and parts list for a test setup that will perform this task.

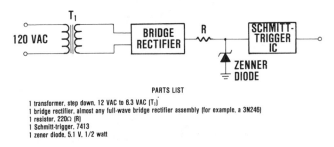

PARTS LIST

1 transformer, step down, 12 VAC to 6.3 VAC (T$_1$)
1 bridge rectifier, almost any full-wave bridge rectifier assembly (for example, a 3N246)
1 resistor, 220Ω (R)
1 Schmitt-trigger, 7413
1 zener diode, 5.1 V, 1/2 watt

Figure 3-28: Test setup and parts list for a test circuit for the 7413 Schmitt-trigger IC.

In order to work with this IC, you need to have at least a nodding acquaintance with the input circuits. Logically, each circuit functions as a 4-input NAND gate but, because of the Schmitt action, the gate has different input threshold levels for positive- and negative-going signals. Figure 3-29 shows a schematic for a single gate input.

The special Schmitt-trigger levels are designated $V_T +$ and $V_T -$. The V_T voltage level is the input level at which a positive-going signal causes the trigger circuit output to switch from logic 1 to logic 0. The V_T- level is the input level at which the negative-going signal causes the output to switch from logic 0 to logic 1 and be ready for the next cycle. There are minimum and maximum voltages for both $V_T +$ and $V_T -$. These are given for the 7413 IC, in Table 3-6.

What happens between the two levels (shown as V_T+ and V_T-, in Table 3-6) in the *hysteresis* interval, depends on the state of the output as the input

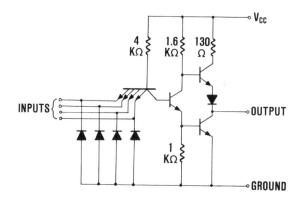

$V_{IN}(1)$ Logical 1 input voltage required at all input terminals to ensure logical 0 level at output

$V_{IN}(0)$ Logical 0 input voltage required at any input terminal to ensure logical 1 level at output

Figure 3-29: Schematic of a single 4-input NAND gate.

PARAMETER		TEST CONDITIONS	MIN	TYPICAL	MAX	UNIT
V_{T+}	POSITIVE-GOING THRESHOLD VOLTAGE	$\overline{V_{CC}} = 5V$	1.5	1.7	2	V
V_{T-}	NEGATIVE-GOING THRESHOLD VOLTAGE	$V_{CC} = 5V$	0.6	0.9	1.1	
$V_{T+} - V_{T-}$ HYSTERESIS		$V_{CC} = 5V$	0.4	0.8		V

Table 3-6: Voltage values of the 7413 IC thresholds designated V_T+ and V_T-.

signal enters the interval. If the output is at logic 1 when your injected input signal enters the hysteresis level, it will stay at logic 1 until your input signal exceeds the V_T+ threshold level. On the other hand, if the output happens to be at logic level 0 during the time your input signal enters the hysteresis interval, you will find that the output will remain at logic 0 until the signal drops to about 0.8 volts (assuming $V_{cc} = 5V$). Those threshold actions are why the Schmitt-trigger is able to clean up noise or irregular signals and why the IC can be used to change the 60 Hz line current to the logic levels produced by the circuit shown in Figure 3-28.

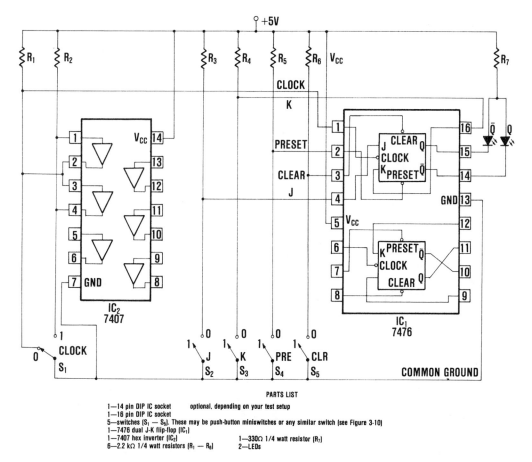

Figure 3-30: Schematic diagram and parts list for testing a flip-flop IC.

How to Use and Test Flip-Flop ICs

Test Equipment:
  See **Comments.**

Test Setup:
  See Figure 3-30.

Comments:
  For this test, we will use the inexpensive 7476 IC. You will also need a hex converter, to construct a trigger pulse bounceless switch. The schematic diagram and parts lists are shown in Figure 3-30. After you have the 7476 and 7407 ICs wired up, you can refer to Table 3-7 for an understanding of what happens when you operate the switches.

PRE	CLR	J	K	CLK	Q	Q̄	MODE
0	1	X	X	X	1	0	ASYNCHRONUS PRESET
1	0	X	X	X	0	1	ASYNCHRONUS CLEAR
1	1	1	0	⊓	1	0	SYNCHRONOUS PRESET
1	1	0	1	⊓	0	1	SYNCHRONOUS CLEAR
1	1	0	0	⊓	Q_{t-1}	$\bar{Q}_{t-1}$	MEMORY
1	1	1	1	⊓	$\bar{Q}_{t-1}$	Q_{t-1}	TOGGLE

PRE=CLR = 0 IS INVALID
X=DON'T CARE=
⊓ COMPLETE BLOCK WAVEFORM

Table 3-7: Truth table for a 7476 dual J-K master-slave flip-flop with preset and clear.

The preset and clear inputs preset the J-K flip-flop to a desired state before another operation is begun. These two inputs are referred to as asynchronous inputs because they do not require a transition on the clock input. You should find that the J and K inputs affect only the Q and Q̄ inputs (LEDs) when you cause a transition to happen (press the clock switch) on the clock input. If the J input is 1 and the K input is 1, the flip-flop should reset from the previous state when you change the clock input from low to high.

Procedure:

Step 1. To set the IC, apply a 1 to the J input and a 0 to the K input, then apply a low-to-high transition to the clock input. This operation is synchronous with the clock operation, i.e., with the clock switch.

Step 2. Put the circuit into a memory mode by setting preset = clear = 1 and J = K = 0. Once you have caused a complete clock pulse to occur, the circuit will remember the states it held at the end of the previous clock pulse.

Step 3. You can operate the flip-flop in a toggle mode (switch back and forth from high to low), as long as you keep preset = clear = J = K = 1. You will find that the negative-going edge of each clock pulse will switch the output state.

One last comment—setting preset = clear = 0 is an invalid operation and you should not use this mode. In general, you should only change the J and K inputs while the clock pulse is at logic 0. In fact, if you try to change the J and K inputs while the clock input switch is in the high position, you will have to cycle the entire operation all over again. You will find that the Q outputs will not show any change until the clock drops to 0 and the circuit is recycled.

Other flip-flops can be tested by using the appropriate truth table and setting up the input switches before the clock pulse is initiated.

Remember, leave the power off on your test setup when installing and removing ICs from the sockets. Also, you will find that an oscilloscope is essential, if you want to observe the actual waveforms.

How to Determine the Duty Cycle of an IC Multivibrator

In a previous part of this section, we worked with the 555 IC timer. Once again, we will use this IC. However, this time we want to determine its duty cycle. Figure 3-31 shows the connections for this test.

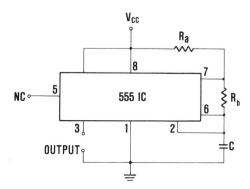

Figure 3-31: Wiring diagram for determining the duty cycle of an IC multivibrator.

If you are in the process of building this circuit on your breadboard, it is best to use a standard value of capacitance for C, then calculate the required resistance. You can always use different values of resistance in series, parallel, or combinations, but, as we all know, it is difficult to find oddball values of capacitance in a spare parts box. For this test, the equation you should use is:

$$\text{duty cycle} = (R_a + R_b)/(R_a + 2R_b) = t_2/t_1$$

As you can see from this equation, we are particularly interested in the two resistors labeled R_a and R_b in Figure 3-31. Why? Because these two resistors are important if you want to predict or design a certain duty cycle, using the 555 timer in an astable circuit.

Of special interest in the formula for duty cycle is the fact that if you make R_a equal to zero, it places pin 7 at V_{cc} potential and the duty cycle equal to 0.5. *Don't do it!* There is no internal current limiting resistor within the 555 IC. Your best bet, in most applications, is always to select a duty cycle of 0.525 or 0.530. To repeat; *the duty cycle of an astable 555 must be greater than 50 percent.*

The circuit shown in Figure 3-31 allows a wide selection of both frequency and duty cycle from a single capacitor. For example, if you use a capacitor value of 0.01 µF and a resistance of 2200 ohms for R_a, you can then vary the value of R_b for several frequencies and duty cycles. For instance, when R_a = 2200 ohms, R_b = 10,000 ohms, the duty cycle = 0.054 or 54%. Setting R_a and R_b both to equal 100,000 ohms, will produce a duty cycle of 200/300, or 66%.

If you use trimmer potentiometers for both R_a and R_b, the frequency and duty cycle can be trimmed to your exact requirements. Just remember: *do* *not set the trimmer R_a to zero resistance.* To be on the safe side, always start with R_a set at a high value. *Note:* there are several different formulas for duty cycle (used by the 555 IC manufacturers, and other authors) for operation of this IC in the astable mode. The equation given in this test has been checked and double-checked, and found to be correct.

SECTION 3.3: CONSUMER ICs

The integrated circuits for consumer applications include all those offered by manufacturers (such as Motorola) for use in equipment that is generally classified as "consumer electronic," i.e., such things as audio preamplifier ICs, sound detector ICs, 1st and 2nd video IF amplifiers, automatic fine tuning, and a host of other video circuit ICs, as well as OP AMPs, phase lock loops, etc. Therefore, if you desire information pertaining to ICs used for consumer applications, refer to the index or table of contents of this book for the particular application you are working with. Incidentally, many consumer ICs are custom-made (for example, read-only memory ICs are frequently programmed to the customer's specifications) and, if replacements are needed, only the original equipment manufacturers have them in stock. Generally, it is very difficult to obtain any publications or other data on consumer ICs such as these.

SECTION 3.4: INTERFACE ICs

Analog-to-Digital/Digital-to-Analog Conversion ICs

One necessary stage in any digital system that processes information originating in analog form is the analog-to-digital (A/D) converter. It is used to convert analog electrical current or voltage levels to representative digital words in digital subsystems. As one would imagine, digital-to-analog conversion is the generation of analog current or voltage levels in response to digital words.

Take, for example, a 3-1/2-digit digital voltmeter (DVM). The instrument's input section is analog (inputs for volts, milliamperes, ohms, etc.), but the output (decoder/driver and display section) is digital. The A/D converter fills the gap between these two subsystems.

Several companies manufacture a set of ICs that can be used to construct an analog-to-digital converter subsystem. One such system uses Motorola's MC1505 and MC14435 ICs. The MC1505 uses the proven dual-ramp A/D conversion technique. The MC14435 is the digital logic section of the converter and is used to produce the complete 3-1/2 digit DVM function.

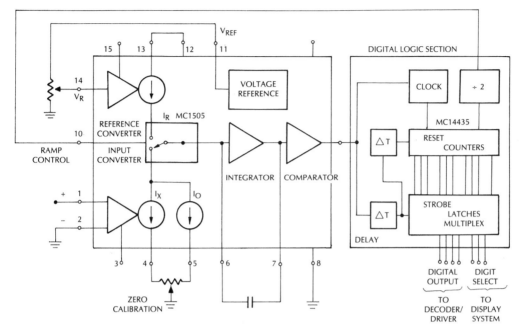

Figure 3-32: Analog-to-digital converter subsystem functional diagram and pin connections, using MC1505 and MC14435 ICs (see Figure 3-33 for MC14435 pin connections).

A functional diagram with pin connections using these two ICs is shown in Figure 3-32.

Due to its simplicity, the dual-ramp (also called *dual-slope*) conversion technique is very popular. Referring to Figure 3-32, the MC1505 (the dual-ramp converter) is an integrating A/D converter in which the analog signal is converted to a proportional time interval, which is then measured digitally. In this system, an integrator (pins 6 and 7) is used as an input to a comparator (output at pin 9). There are, also, an on-chip voltage reference, a pair of voltage/current converters, a current switch, and associated control and calibration circuitry. Only two capacitors and two calibration potentiometers, zero calibration (pins 4 and 5), and full-scale calibration (pin 11 and ground), plus V_{cc} (+16.5 Vdc *maximum rating* for the MC1505) and ground, are required for breadboarding this pair of ICs. The *maximum* V_{cc} rating for the MC14435 is + 18 Vdc. See Figure 3-33 for the pin connections and functional block diagram of this IC.

The A/D converter IC is also used to convert analog currents and voltages to represent digital words for use in computers and microprocessors. In fact, now that MCUs and MPUs are so widely used, A/D converters have become even more popular than they were when most were used in control

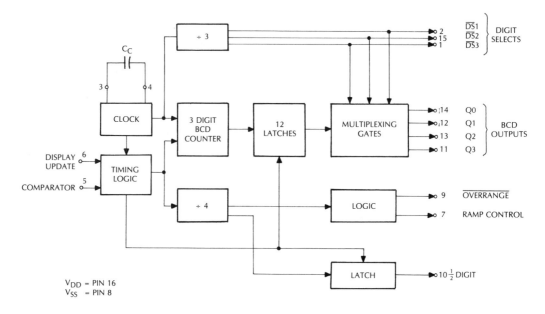

NOTE: MC1505/1405 A/D Converter Subsystem recommended for linear front end.

Figure 3-33: Block diagram and pin numbers for the MC14435. A 3 1/2 digit A/ D logic subsystem (see Figure 3-32 for the complete converter subsystem).

systems, instruments, and the like. But these A/D converters are not of any use until they are interfaced to a MPU or MCU.

There are several approaches to this job. First, in some cases (such as in the design of the MC68705R3, an 8-bit **EPROM** microcomputer unit), the 8-bit A/D converter is implemented on the chip. Up to four external analog inputs (via port D, in this case) are connected to the A/D through a multiplexer. Second, if you have a MCU, it may require only that you plug in an I/O cable to the input slug on the back of your computer. But, if you are breadboarding, you may have to interface directly to the MPU or MCU, and this requires a knowledge of the MPU control signals (see Sections 3.5 and 3.6).

As in MPU interfacing jobs, there are two approaches that you can take: the I/O route or the memory-mapped way. Section 2.7 explained the I/O based system. It is possible that you may have to interface with the bus lines inside your computer, or directly to the MPU. In any event, you will have to provide some means to identify the control signals and decode the I/O addresses.

In the memory-mapped systems, the output of the A/D converter is seen as another location in memory. Or, to put it another way, the term "memory-mapping" is telling you that the CPU will see the A/D converter as a memory location.

Peripheral Interface Using Drivers and Receivers in Computer Applications

There are many line drivers and receivers for computer/terminal applications. You will usually find them listed in pairs in sales catalogs (a driver and a receiver). For example, the MC11488 IC is a *quad driver* with output current limiting and the MC1489 is a *quad receiver* with an input voltage range of ± 30 volts. Figure 3-34 illustrates a typical application of this pair of ICs. Incidentally, note that the input shows MDTL (DTL is an abbreviation for diode transistor logic and the M stands for Motorola). Also, this driver is compatible with all Motorola MTTL logic families.

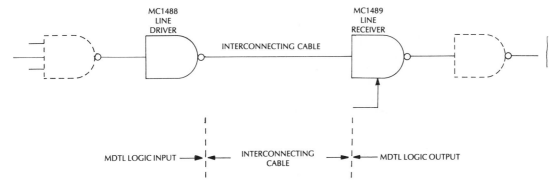

Figure 3-34: Typical application using Motorola quad driver and receiver interface ICs.

It is also possible to adjust the input threshold voltage for the MC1489 receiver IC. This is particularly important because it makes possible excellent interfacing between MOS circuits and MDTL/MTTL logic systems. Figure 3-35 shows how this may be done by placing an external resistor (R) in the IC circuit.

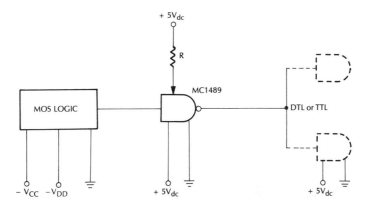

Figure 3-35: Using an MC1489 to interface MOS logic to DTL or TTL logic.

In this application, the input threshold voltage is adjusted by using a certain voltage (to the external response control pin) and a predetermined resistor value that must be adjusted (the voltage and resistor values), until the threshold voltage falls to about the center of the MOS device's voltage logic level. Figure 3-36 illustrates the input threshold voltage adjustment for one of

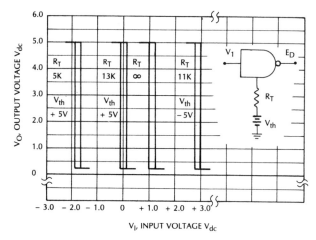

Figure 3-36: Input threshold voltage adjustment chart for MC1489 line receiver.

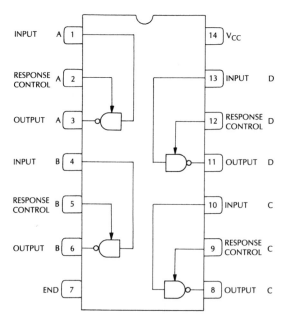

Figure 3-37: Logic diagram and pin configuration for the MC1489L quad line receivers.

the receivers in the quad receiver. Each receiver has one output terminal, one response control terminal, and one input terminal (see Figure 3-37).

Working with Bus Interfaces

A bus in a computer is usually 8, 16, or 32 lines used as a path over which information is transmitted from any of several sources to distinct destinations. You may see the term *highway* used when a bus interconnects many system components. Also, you will encounter the term *handshake bus*. This is used to indicate a bus that is used for connecting the MPU-based system to some peripheral.

If you are servicing a system using a standard 8080 or M6800 MPU, the maximum number of ICs you will find connected to it (without using data bus extenders) is 3 RAMs, 3 PIAs, and 1 ACIA. Figure 3-38 shows the logic diagram and pin numbers for Motorola's MC6880 data bus extender.

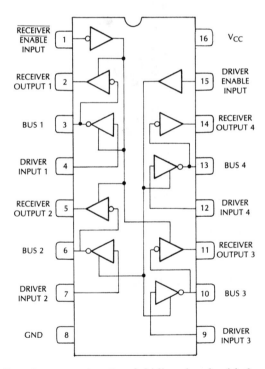

Figure 3-38: Data bus extender. Quad, bidirectional, with 3-state outputs.

This family of ICs (the MC6880 is an inverting type and the MC66889 is a non-inverting type) is designed to extend the limited drive capabilities of the NMOS type 6880 and 8080 MPUs. The maximum input current of 200 μA at any of the data bus extender input pins assures proper operation

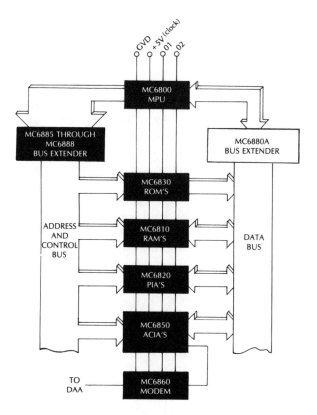

Figure 3-39: MPU bus extender application.

despite the limited drive capability of the MPU chip. Figure 3-39 shows a MPU bus extender application using Motorola's components you might encounter at your workbench.

SECTION 3.5: SERVICING TODAY'S COMPUTER-ON-A-CHIP ICs

Computers-on-a-chip ... what are they? A single *microprocessor* unit such as Motorola's MC6800 is an integrated circuit that performs many of the functions, such as arithmetic and control, usually found in a digital computer. However, the MC6800 by itself does not contain the memories (RAMs and ROMs) and input/output (I/O) functions of a computer. Therefore, such an IC is called a *microprocessor* (MPU), not a *microcomputer* (MCU). But, add the necessary memories and I/O functions to a system and now a microcomputer unit is formed. Today there are many ICs that contain all these functions. In effect, when an IC contains all the basic functions, it is a computer-on-a-chip.

To review briefly, in most computer systems there is a read only memory to store the computer instructions, and random access memory to store temporary data (the data to be manipulated by the computer program), a microprocessor, and an I/O integrated circuit to make the system interface with outside or peripheral equipment (intersystem communications). For example, video terminals, line printers, keyboards, etc. are required. Include all these functions on a single chip and you have a computer-on-a-chip, or MCU.

One 8-bit microcomputer unit is Motorola's MC6805R2. This unit contains a microprocessor (referred to as a *Control Processor Unit,* or CPU, which is another name for microprocessor), on-chip clock, ROM, RAM, I/O, 4-channel 8-bit analog-to-digital (A/D) converter, and timer. This IC features 64 *bytes* of RAM and 2048 bytes of user ROM. *Note:* A byte is a unit composed of

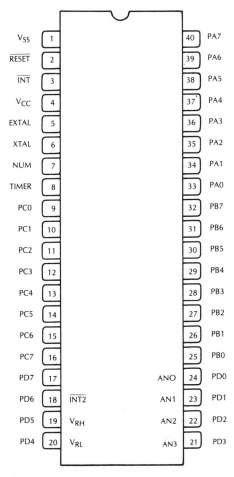

Figure 3-40: 8-bit microcomputer unit (MC6805R2) pin assignments.

eight binary digits (8 bits). Figure 3-40 shows the pin assignment for the MC6805R2.

All the circuit elements of the MCU are in integrated circuit form except, of course, external wiring, readouts, keyboards, and the like. Thus, you do not have access to the circuit elements, nor can you change them in the way they function.

It should be pointed out that you need not understand every detail of the internal circuits of this IC (or any similar one) to effectively use and service MCUs. Nevertheless, you do have to know what the MPU registers and counters are and how (in basic terms) they operate.

Understanding Microcomputer Address, Data, and Control Lines

By now you are somewhat familiar with the RAM, ROM, I/O, and MPU ICs, but we are now interested in addressing and controlling these devices. Figure 3-41 is a basic microcomputer system.

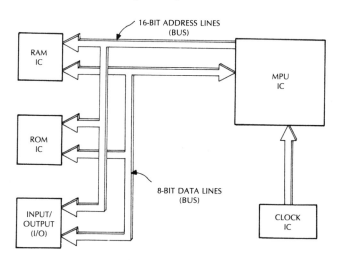

Figure 3-41: Basic microcomputer system.

It is important to understand that the MPU will do only what the program in the computer directs it to do. These instructions are sent over connecting wires (called *lines*) between the MPU, RAM, and ROM. Collectively, these lines are usually called *bus, address bus,* or *data bus.* See Figure 3-41.

Some MPUs, today (such as the MC6802, etc.), contain 16-bit memory addressing (A15......A0). An address is generated using a binary code—logic highs and logic lows. As you may know, a group of 8 bits is 1 byte, which means that an address 16 bits wide would equal 2 bytes wide. The other lines (8-bit data lines) shown in Figure 3-41 are also typical in modern MPUs. As the name "data" suggests, these lines (usually labeled D6.....D0) are used to send and receive data; therefore, data bus. Usually, data bus is bidirectional between memory and peripheral devices. Again, data is in the form of binary words, i.e., high logic and low logic.

There is a third type of input to an MPU. This is called a *control bus*. These lines are used to control the order of operation of the entire system. For example, it may be desired to write into memory or, on the other hand, it may be that the function is a read only operation. Some of the control bus lines on a MPU are READ/WRITE (R/W), bus available (BA), HALT, IRQ, RESET, NMI, RE, VMA, MR, and (E). There are two other connections found on the MC6802: the crystal connections (EXTAL and XTAL). See Figure 3-42 for the pin assignments on the MC6802 MPU with clock and 128 bytes of on-board RAM.

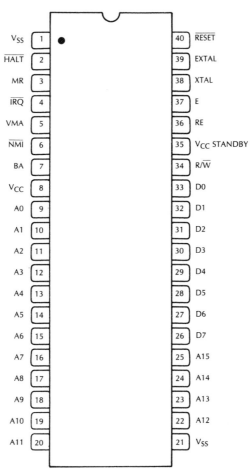

Figure 3-42: Pin assignment for a MC6802 microprocessor. See text for MPU signal descriptions.

Frequently Needed MPU Signal Descriptions

Just as the MPU must address the device that it wants to communicate with (using address pins A0...A15), it also must have some channel through which it can send data, or receive data, as has been explained. This is done by using DATA BUS pins (D0.....D7). The other pins shown in Figure 3-42 are basically control lines (except V_{cc} and V_{ss}, of course). A brief description of each of these various operations follows. See Figure 3-43 for a block diagram of a *typical* MPU using a MC6802.

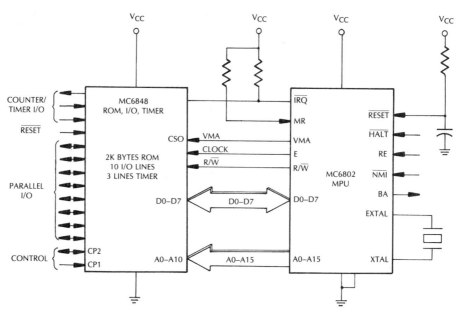

Figure 3-43: Basic microcomputer designed using an MC6802 MPU and a MC6848 ROM, I/O, timer. This minimum component system may be expanded with other parts of the M6800 microcomputer family.

Halt: When this input (pin 2) is in the logic 0 state, all activity in the machine will halt.

Read/$\overline{\text{Write}}$ (R/$\overline{\text{W}}$): This output (pin 34) informs the peripherals and memory devices whether the MPU is in a read (1) or write (0) state.

Valid Memory Address (VMA): This output (pin 5) signals peripheral devices that there is a valid address on the address bus.

Bus Available (BA): This signal (pin 7) will indicate when the microprocessor has stopped and the address bus is available, and when it is activated, i.e., goes to a logic 1 state from its normal logic 0 state.

Interrupt Request ($\overline{\text{IRQ}}$): A logic 0 on this input (pin 4) requests that an interrupt sequence be generated within the machine.

Reset: This input (pin 40) is used to reset and start the MPU when a power failure occurs, or during initial startup.

Non-maskable Interrupt ($\overline{\text{NMI}}$): A logic 0 on this pin (pin 6) requests that an interrupt sequence be generated within the MPU.

In summary, as we have stressed, all MPU, ROM, EPROM, etc. operate with binary numbers only. However, most manufacturers do define a mnemonic code (usually two or three letters of the alphabet) that describes the function of each instruction. Some mnemonic codes and operations used by Motorola are shown in Table 3-8. For example, ABA (mnemonic code) is the instruction for the MPU to add accumulator B to A. This instruction would be stored in ROM memory in binary form, i.e., 1B = 0001 1011 in memory.

MNEMONIC	HEX	OPERATION
ADDA	9B	ADD
ABA	1B	ADD ACCUMULATOR
CLR	4F	CLEAR
CMPA	9I	COMPARE
ASL		SHIFT LEFT ARITHMETIC
LSR		SHIFT RIGHT, LOGIC

Table 3-8: Mnemonics.

How and Why Addressing Modes Are Used in a Microcomputer System

Instructions are executed in sequence, step by step, by an MPU, unless it is told to follow some other instruction in a different location in memory. As the MPU performs whatever instruction it has received, it is often necessary to have data. The needed data is generally located in memory, and sometimes is found in more than one memory location.

Memory locations in MCUs such as the MC6805R2 are 8 binary digits wide. Incidentally, there are many 8-bit chips; for instance, the EPROM microcomputer unit MC68705U3, the microcomputer unit MC6805P4, and a host of other MPUs, memory, etc. However, there are also 4-bit single-chip MPUs such as the CMOS MC141000 and, of course, some of the newer chips will handle 32 bits.

As you would guess, instructions in an 8-bit MCU memory are 8 bits wide (1 byte) and the instructions in a 4-bit system are 4 bits wide. When an instruction requires the use of a certain memory location and one or two following locations, it may be referred to as a 2-byte, 3-byte, etc. instruction.

Memories are divided into locations called *addresses* and the contents of these addresses can be instruction, data, or a combination of both. There are several methods for arranging the data bytes in memory. The MCU we have been talking about (the 8-bit MC6805R2) has 10 addressing modes available for use by the programmer. But MPUs (the heart of an MCU) vary in design, with each programmable only bit having its own set of instructions.

In general, each family of processors (8080, 8040, Z80, and 8085, or the M6800, MC6809, and all 6800 family MCUs) require a certain internal programming language. Nevertheless, most information (*but not all*) will apply to the entire family of the particular manufacturer. Instructions not used by Motorola's 6800 family will not, however, be covered.

Before we discuss addressing modes, you will need to understand the on-chip registers and accumulators available in the MPUs. *Note:* In most instances, *accumulator* is simply another word for *register.*

There are several generation additions to the M6800 family that have architectural improvements (arrangement of counters, registers, arithmetic logic units, and so forth, within the MPU) which include additional registers, instructions and addressing modes, but, to make it as simple as possible for the moment, we shall first consider only the M6800 MPU registers and accumulators:

A and B Accumulator: The MPU contains two 8-bit accumulators which are used to hold operands and results from an arithmetic logic unit (ALU).

Index Register (IR): This register contains 2 bytes, i.e., a 2-byte register. It is used to store data or a 16-bit memory address for the index mode (to be explained later) of memory addressing.

Program Counter: This counter is a 2-byte (16-bit) register that points to the current program address. In other words, it contains the address of the next instruction to be fetched from memory. Early in the execution phase of each instruction, the program counter is automatically advanced to the address of the next sequential instruction in memory.

Stack Pointer (SP): Most processors (for example, the 8080 family or M6800 family) have a special method of handling subroutines to insure an orderly return to the main program. This is needed when there is a program within a program. The stack (a 16-bit register) holds the memory address of the instruction to be executed after the subroutine is completed. The stack is normally a Read/Write memory that may have any location (address) that is convenient. It should be pointed out that various MPUs have different methods of maintaining their stack. A *stack pointer is an internal register of the CPU.*

Condition Code Register (CC): The condition code register indicates the results of an arithmetic logic unit operation. It is a 1-byte register that is used to test the results of certain instructions (this will be explained in detail later).

Now that you have a basic understanding of the registers and accumulators in an MPU, you will also need a brief description of the various addressing modes. This is one of the other basic features you must comprehend before servicing equipment containing an MPU. "Addressing modes" refer to the manner in which you cause the MPU to obtain its instructions and data. And, as we have shown in previous pages, you must have a method for addressing the MPUs on-chip registers and all external memory locations. The fundamental addressing modes for the M6800 family follow:

Inherent Addressing mode: The instructions in this mode are always in 8-bit form and do not require an address in memory. An example of an inherent instruction is the mnemonic instruction "ABA." This instruction tells the MPU to "add the contents of accumulators A and B together and place the result in accumulator A."

Accumulator Addressing Mode: Like the inherent mode, the accumulator mode is also an 8-bit instruction. This instruction is directed to either the A or the B accumulator. For example, the mnemonic instruction 1NCB causes the contents of accumulator B to be increased by one. This is an increment operation and the Boolean/arithmetic operation is written as B = 1 →B. *Example:* COMB (complement the B accumulator) instruction; after execution of the COMB instruction, each of the eight bits in the B register will have been inverted, i.e., all 1s become 0s and all 0s become 1s (Boolean/arithmetic operation B →B).

Immediate Addressing Mode: This addressing mode is used for operations such as "Load accumulator A," mnemonic LDAA (Boolean M − >A). For example:

Operator	*Operand*	*Comment*
LDAA	#20	load 20 into ACCA

This instruction causes the MPU to immediately load accumulator A with the value 20. Notice, the *operand* is the value (number 20, in our example) to be operated on.

Direct Addressing Mode: Direct addressing generates a single 8-bit operand, and hence can address only memory locations 0 through 255 in our example chip, the M6800 (this chip has extended address, which permits use of more locations, but we will come to that in the next mode). In this mode, the *address* where the data is located is in the next memory location.

The MPU, after encountering the operation code (op-code) for instruction Load Accumulator A (LDAA, direct) at a certain memory location, looks in the next location for the address of the operand. It then sets the program counter equal to the value found there and fetches the operand (in our example, a value to be loaded into accumulator A) from that location. Because the address of the data must be specified in 8 bits, the lowest direct address is 000 000, and the highest is 111 111 (hex-FF).

Extended Addressing Mode: The direct and extended addressing modes differ only in the range of memory locations to which they can direct the MPU. As was pointed out in the direct mode, only address 0 through 255 can be used with the M6800. A 2-byte operand is generated using this MPU for extended addressing, enabling the M6800 to reach the remaining locations 256 through 65535. The instructions of the direct addressing mode are 2 bytes wide and the extender mode instructions are 3 bytes wide. One memory location is used for instruction and the next two for data location, in the extended mode.

Extended address is a method of reaching any place in memory. However, direct addressing is a faster method of processing data because, in most applications, it uses fewer bytes of control code.

Relative Address Mode: In the last modes, we discussed direct and extended. The address obtained by the M6800 is an absolute numerical address located immediately following memory location and, in extended, the next two memory locations.

In the relative addressing mode, the next instruction to be executed by the MPU is located at some memory address other than the one following. This mode of operation is implemented for the MPU's *branch* instructions. These instructions specify a memory location relative to the program counter's current location. Branch instructions generate 2 bytes of machine code: one for the "instructions opcode" and one for the "relative" address.

The M6800 is capable of branching both forward and backward. The next location in memory after the branch instruction contains the information that tells the MPU how to branch (branch ahead or branch back from its present location) for its next instruction. The MPU will then continue executing instructions from the new location in memory.

Indexed Addressing Mode: With this mode, the numerical address is variable and is dependent upon the real time contents of the index register. This mode of addressing uses a 2-byte instruction, in which an instruction followed by a number that is added to the contents of the index register to form an address, or source statement such as

Operator	*Operand*	*Comment*
STAA (store accumulator)	X	Put A in indexed location

causes the MPU to store the contents of accumulator A in the memory location specified by the contents of the index register. It should be mentioned that the level X is reserved to designate the index register, when working with the MC6800 MPU. For example, the TSX instruction for this MPU causes the index register to be loaded with the address of the last data byte put onto the "stack."

Summary

An MPU signal description (pin by pin) was given in this section. The previous pages described the various accumulators and registers and the different addressing modes used to load them. The most important point presented is that if, while you are making up a program, the actual data is in the next one or two bytes, you should use the instruction code for the immediate mode. Or, if the data is one byte wide, use the code for the direct mode. On the other hand, if the data is located at an address that requires two bytes to describe, you should then use the instruction code for the extended mode. That is to say that the instruction to load A or B register can require any of several different modes, and the hex code for each mode is different, therefore the binary code will be different. Referring to Table 3-9, you will see that it is absolutely critical that you consult the manufacturer's instruction sheet before programming a specific MPU IC.

MODE	INSTRUCTION CODE		
LOAD A IMMEDIATE	86	1000 0110	← BINARY
		8 6	← HEX
LOAD A DIRECT	96	1001 0110	← BINARY
		9 6	← HEX
LOAD A EXTENDED	B6	1011 0110	← BINARY
		B 6	← HEX
LOAD A INDEXED	A6	1010 10110	← BINARY
		A 6	← HEX

Table 3-9: Addressing modes to load A register, showing mnemonic instruction code (hex) and binary code.

SECTION 3.6: PRACTICAL GUIDE TO SERVICING MICROCOMPUTER ICs

Whatever your prime interest in a microcomputer is, be it system building, maintenance and service, programming or experimenting with the various chips, you should familiarize yourself with modern on-chip microcomputer units. Microcomputers vary in design, with each design programmable only through its own set of instructions, as has been explained. Some of the MCUs and MPUs we will cover in this section are the MC6805 MCU family and the MC68000 MPU family.

This section is primarily devoted to MCU operation and it is assumed that the reader has read Section 3.5 and gained a knowledge of MPUs. If you have studied MPU basics thoroughly, you should have no difficulty in understanding the diagrams and descriptions presented in this section.

All examples used in the following pages were obtained from actual MCU literature provided for the author by the Motorola Corporation Semiconductor Division, located at 3501 Ed Bluestein Blvd., Austin, TX 78721. However, the basic approaches to using MCU chips presented here can be applied to almost any microprocessor/stored-program system now being developed (much of the information in this section is advance information), to those that will be manufactured in the future, as well as to all the MPU-based ICs now in use. *Note:* This statement does not apply to some of the new computer applications such as artificial intelligence, and the like.

Servicing Today's Microcomputer Units

As we indicated in Section 3.5, the MPU is the heart of the system. But in order to compose a computer system, support circuitry is required. As you will remember, in addition to ROM (instruction memory) there is also a need for working data storage (RAM), and the MPU must provide the required timing, plus control signals. Of course, I/O data instructions and addresses are also necessary. One example of today's MCU units is the MC6805P2. This computer-on-a-chip contains a CPU (central processing unit), on-board clock, ROM, RAM, I/O, and timer. Figure 3-44 shows a block diagram of this 8-bit MCU unit. This MCU comes in a 28-pin package that is illustrated in Figure 3-45.

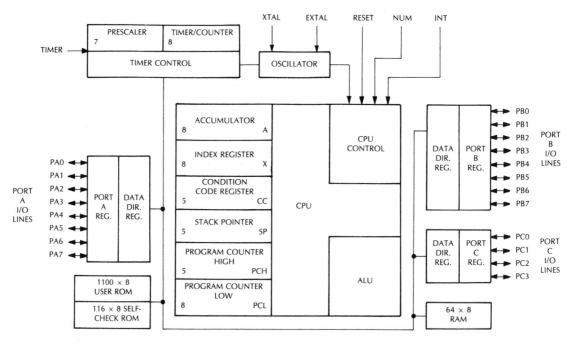

Figure 3-44: A state-of-the-art MCU block diagram (MC6805).

An important point is that you, as a technician, realize that once you understand the input and output signals of an MCU family, you'll find one unit is very similar to another. For instance, refer to Figure 3-40 in Section 3.5 (pin assignments for the MC6805R2), and it should be apparent there is not much input signal difference except that the MC6805R2 has more user ROM (2048 bytes versus 1100 bytes), more I/O lines (24 versus 20), etc., and, of course, 40 pins, whereas the MC6805P2 is a 28-pin IC.

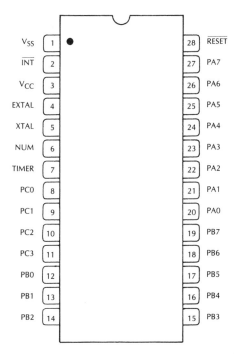

Figure 3-45: Package drawing and pin assignments for the 8-bit MCU MC6805P2.

Because the MC6805P2 is a simpler device, let's examine its block diagram (see Figure 3-44) and see just what its input and output signals are.

V_{cc} **and** V_{ss}**:** Notice that pins 1 and 3 are V_{ss} and V_{cc} respectively. This is not true of all MPUs. For example, the MC68701 uses pins 1 and 7 for V_{ss} and V_{cc}, the MC6805P4 uses pins 1 and 3, but the MC6800 and MC68705R3 use pin 1 (V_{ss}) and pin 4 (V_{cc}), as does the MC6805P2 we are discussing.

$\overline{\text{Int}}$**:** This pin feeds the CPU control (see block diagram, Figure 3-44) and provides a way of applying an external interrupt to the MCU. Although different MCUs use different pins for the input, they all perform the same basic task. However, MCUs have three different ways in which they can be interrupted (as has the MC6805P2), and some have four ways—the MC6805R2, for example. Regardless of what method is used for interrupt, when any interrupt occurs, processing is suspended.

XTAL and EXTAL: The block diagram (Figure 3-44) shows these pins (Figure 3-45) connected to the on-board oscillator (clock circuit). If you are working with the MC6805P2, a crystal (usually 4.0 MHz), a resistor, a jumper wire, or an external signal may be used to generate a system clock. With other MCUs, sometimes you can use only a crystal or resistor. For instance, the MC6805R2 requires an "AT" cut crystal at a maximum frequency of 4.0 MHz.

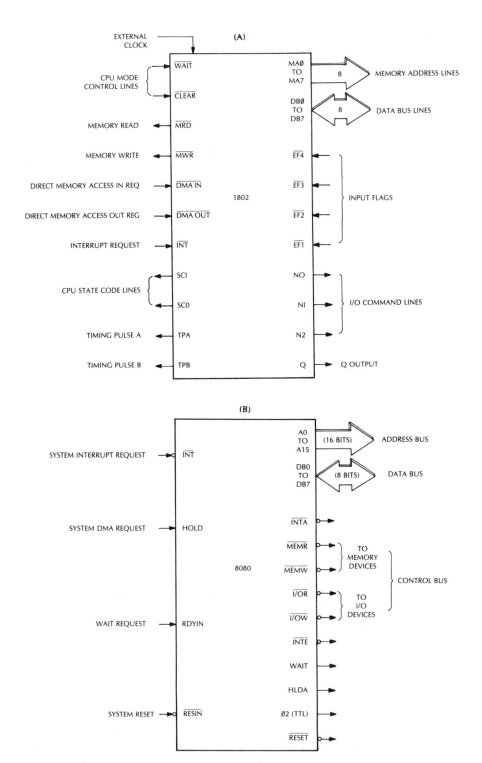

Figure 3-46: Inputs and outputs for the 1802 (A) and the 8080 (B).

Timer: This input (pin 7, Figures 3-44 and 3-45) allows you to use an external input to decrement the internal timer circuitry. More about this function will be presented later in this section.

Reset: This pin is tied to the CPU control, as shown in Figure 3-44. Basically, you can reset the MCU at times other than those automatically used with the circuits built into the MCU.

NUM: You, as an MCU serviceman, do not use this pin (pin 6, Figure 3-45). It must be connected to V_{ss} (pin 1, Figure 3-45). All other lines such as PA0 ... PA7, and PC0 ... PC3 (Figure 3-44) are I/O ports, as shown.

Other MCUs may have more or fewer I/O lines, accumulators, and registers (this one has five registers available to the programmer), and more or fewer RAM (64 bytes, in this case), ROM (1100 × 8 user ROM and 116 × 8 self-check ROM, shown in Figure 3-44) but, as we have explained, in general all the MCUs in a certain family will follow the same number-bit and architecture, use basically the same control operations, and, in the case of the M6805 family, have software compatibility similar to that of the M6800 8-bit microprocessing unit. In fact, this is usually true of all MCUs. Whatever the basic MPU is—for instance, the 8080—the entire MCU family will be developed using similar hardware and software.

As an example of the similarity between inputs and outputs, let's compare the MC6805P2 with the 1802 and 8080. We have explained all the inputs and outputs of the MC6805P2 in the last few pages, so all we need now is the other inputs and outputs. See Figure 3-46.

Notice that there are differences in signal descriptions, but not so many that you can't quickly grasp what is going on once you have a thorough understanding of one of the various families of MPUs. Nevertheless, it should be apparent that you must read the manufacturer's instruction sheet because MPUs and MCUs vary greatly in terms of their capabilities and specific language or instruction sets. However, as we are trying to point out, they are similar in many functions.

Working with the M6805 Family of MCUs

To service an MCU for any of a host of applications, you need only become familiar with:

1. The microcomputer signal descriptions.
2. The program language.
3. The manufacturer's mnemonics and equivalent hex (or binary).

This requires learning about one hundred new terms and abbreviations (you know most of them), such as the names of the various addressing modes, interrupts, and I/O, MPU, MCU, ROM, RAM, etc.

Four of Motorola's 6805 family are the MC6805P2 (which we have discussed, see Figure 3-44), the MC6805P4, the MC6805R2 (Figure 3-44 shows the pin assignments), and the MC146805G2. We will now examine this family in order to increase your knowledge of MCUs.

First, you already have a fair understanding of the MC6805P2, so let's start off with the MC6805P4, which we have not examined. This MCU is an 8-bit unit that contains a CPU, on-chip clock, ROM, RAM, I/O, and timer. Table 3-10 shows a comparison of the hardware features of the MC6805P4 and MC6805P2.

HARDWARE FEATURES	
MP6805P4	MP6805P2
8-BIT ARCHITECTURE	SAME
112 BYTES OF STANDBY RAM	64 BYTES
STANDBY RAM POWER PIN	NONE
MEMORY MAPPED I/O	SAME
1100 BYTES USER ROM	SAME
20 I/O LINES	SAME
ON-CHIP CLOCK	SAME
SELF-CHECK MODE	SAME
MASTER RESET	SAME
5V SINGLE SUPPLY V_{CC}	SAME

Table 3-10: Comparison of the hardware features of two MC6805 family MCUs.

As you can see from Table 3-10, both these MCUs are very similar in both hardware and software features. At this point, you may be wondering if there are any great differences in the 6800 family of MCUs. To answer this, notice that the MC6805P4 has 112 bytes of RAM, whereas the MC6805P2 has only 64 bytes of RAM. A RAM is an area where data can be stored. This data may be erased at any time and new data stored in its place—obviously very important to the user.

Another MCU, the MC6805R2, has a greater amount of user ROM—2048 bytes. These two MCUs have only 1100 bytes of user ROM. There are several other differences between the MC6805R2 and the two other MCUs we have been discussing. For example, this MCU unit has 24 TTL/CMOS compatible I/O lines. *Note:* All three devices' I/O lines are TTL/CMOS compatible, with eight of these lines being LED-compatible. Also, this IC can be interrupted four different ways, whereas the other two have only three (more about interrupts later); in addition, this unit has an internal A/D converter. To increase your ability to service these and other MCUs, let's now examine each of their internal components in greater detail.

How the RAMs Operate

For our example of how RAMs operate (see Section 2.13 for test procedure), we will assume that there are 128 bytes of RAM and that the MPU has the capability to work with any bit in RAM. To begin, a byte in our example is 8-bits; hence, we are talking about a total of 1,024 memory cells. Because an illustration of that many memory cells would be difficult (and

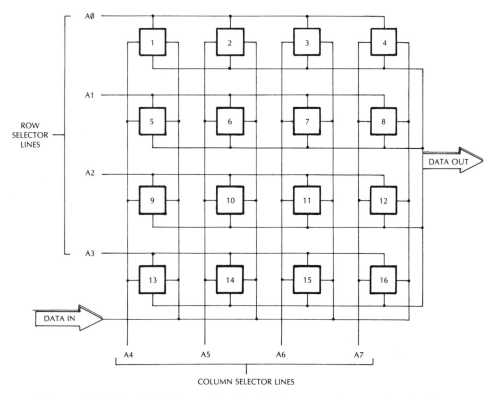

Figure 3-47: Simplified illustration of a 16-bit memory that requires 8 address lines to select any of the memory cells.

serve no real purpose), we will use a simplified drawing of a 16-bit memory array. This is shown in Figure 3-47.

A computer memory is formed from a large array of semiconductor elements, each capable of storing a single binary 1 or 0, usually organized into groups of bits (in most of our examples 8-bits, or a byte). In the example shown in Figure 3-47 each memory cell has a row selector line (A0 ... A7) connected to it. By selecting one of the A0-A3 lines and one of the A4-A7 lines, the desired memory cell is accessed by an MPU and conditioned either to write in (store) a bit or read out a bit. For instance, if Row Select line A2 and Column Select line A5 go to a logic high, memory cell 10 is selected. If the MPU reads only the contents of cell 10, its output will be either 1 or 0, depending on what's stored in the memory cell. The MPU can write a 1 or 0 into memory cell 10 if it places the proper control data input on the Data In bus.

With this simple example, the MPU needs only 8 lines (A0 ... A7) to select any of the 16 memory cells. However, this basic memory cell arrangement restricts the MPU to reading in or reading out only one bit at a time.

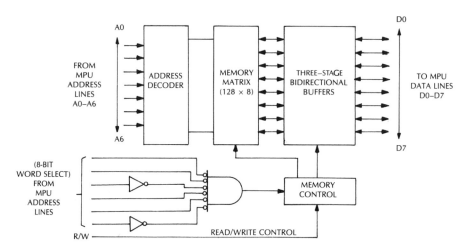

Figure 3-48: Function block diagram of a RAM that can store 128 bytes, i.e., 128 8-bit words.

Figure 3-48 shows the function diagram of a practical modern RAM that is compatible with Motorola's family of 6800s.

The eight RAM data bus pins D0 through D7 are tied to MPU data bus pins D0 through D7. Address pins A0 through A6 are tied to MPU address pins A0 through A6. These seven address pins are used by the MPU to select an 8-bit word on a particular chip addressed.

If the equipment includes more than one RAM in a discrete component system, all A0's would be tied to the A0 address line of the MPU, all A1's would be tied to the A1 address line from the MPU, and so on. Incidentally, using this same line of thinking (adding more memory), it should be apparent that the basic memory cell array shown in Figure 3-47 can be extended to include many more cells: i.e., add additional row selector lines, column select control lines, and memory cells.

When using more than one RAM in a system, there must be some provision to select a certain RAM. You will find that many RAMs have chip select pins. For example, six might be available. Whatever the number, the chip select pins are tied to the address bus from the MPU in such a way that only one RAM chip is addressed at a time.

To illustrate what would happen if you tied three 128-byte RAMs to the same MPU:

128 8-bit words × 3 = 384 8-bit words, or 3,072 individual memory locations.

But, in practical applications, your actual storage is only 384 locations because readout is in 8-bit words.

Practical Guide to Other IC Memories

As has been explained (see Section 2.13 for test procedure), the basic ROM may be programmed by the manufacturer and, in some instances, by the user. In fact, for most systems, numerous ROM programmers are available. A field-programmable read only memory (FPROM) is a memory that you can program. An example of such a device is Signetic's 8223, 256-bit bipolar PROM (organized 32 × 8-bit word). Still better, the erasable programmable read only memory (EPROM) is also available. National Semiconductor Corporation has a 2048-bit (MM5203) that is a 16-pin IC with a quartz window that is transparent to ultraviolet light (see Figure 2-69 in Section 2.13).

Certain applications call for a higher performance system, faster than TTL memories. Motorola Emitter-Coupled Logic (MECL) circuits could very well be used in these cases. They offer a 32 × 9 FPROM (15ns), the MCM10549L ceramic DIP, and a 256 × 4 FPROM (24ns), the MCM10549L. See Figure 2-4 in Section 2.1 for an illustration of a MECL basic gate.

In the MCU category, we find 1-, 4-, 8-, 12-, 16-, and 32-bit processors. Of course this means that a ROM for each MPU must be constructed using the same architecture, i.e., an 8-bit MPU requires an 8-bit ROM, etc. In a micro-programmed MPU, an instruction (from main memory RAM) causes the program memory (ROM) to generate a bit pattern which, in turn, initiates the binary signal (control) that activates the circuit for execution of the instruction.

The type of memory utilized to store these bit patterns may be any of the ROMs we have discussed up to this point. You will find that this group of bit patterns (instructions) that inform the MPU what operation to perform (often called the "instruction set") are usually produced by MPU manufacturers such as Fairchild, Intel, Intersil, MOS, Motorola, National Semiconductor, RCA, Rockwell, Signetics, Texas Instruments, and Zilog.

As an example of a *mask-programmable* (the chip selects, etc. are manufactured in the device) ROM, we will use the MCM6830 that is compatible with TTL, contains 1024 bytes, and is a 24-pin IC. This ROM is much like the RAM circuit shown in Figure 3-47 (see Figure 3-49). But since data cannot be stored in a ROM, a ROM must be addressed solely to obtain data.

This custom-programmed ROM is a 3-state chip. Three-state control is a technique that permits an MPU to see a certain device (in this case, a ROM) as having three states — 1, 0, and an open circuit (high impedance). When the ROM is not addressed by the MPU, it goes into the third state; thus it is possible to shut the ROM off from the system by use of the control inputs to the device. Notice that the user can define whether the chip-select inputs are active high or active low. Also note that the user must define the binary word to be stored at each address. This is true, of course, regardless of whether an FPROM, an EPROM, or a PROM is used.

There are two obvious reasons why a technician would usually want to

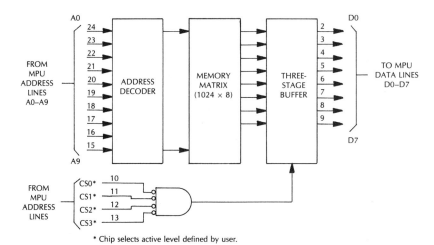

Figure 3-49: Functional block diagram of Motorola's MCM6830 mask-programmable (certain signals defined for the manufacturer, by the user) ROM.

use an EPROM instead of a mask-programmable ROM. First is the cost of having a manufacturer do the programming, and second, and perhaps more important, the technician can work up his program and then try it. If it isn't correct, the ROM is erasable.

However, whatever type ROM is being replaced, tested, etc., it must match the MPU interface lines. If the ROM address lines are A0 ... A9, then they must be connected to the MPU's A0 ... A9 lines. Also, assuming that the ROM has data bus lines, they must be connected to the MPU's D0 ... D7 lines. Furthermore, the chip-enable inputs have to be defined—active high or active low—for the different types of MPUs. As a general rule, you will find more address lines are used for addressing internal locations in ROM in a system than are used for internal address of RAM; for instance, seven lines for RAM and, perhaps, ten lines for ROM.

Input/Output Signals

We have shown that a microprocessor and memories can be manufactured to form a simple and almost complete microcomputer (see Figure 3-44). Incidentally, in many instances, you will find the ROMs and RAMs are separate ICs connected to an MPU and, in other cases where more memory is desired, connected within an MCU system. Notice, we said "almost complete." The missing element is the hardware needed to communicate with the microcomputer system. The MCU usually has an internal I/O integrated circuit to make the system interface with the outside world. Without some I/O hardware, it would be impossible to enter data into or out of an MCU system.

Input/Output Ports

By referring to Figure 3-44, you will find that the MC6805P2 has 20 input/output lines (A0 ... A7, B0... B7, C0 ... C3). Figure 3-50 is an illustration showing typical port I/O circuitry found in MCU's such as this one.

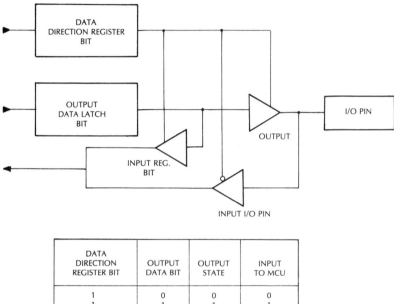

DATA DIRECTION REGISTER BIT	OUTPUT DATA BIT	OUTPUT STATE	INPUT TO MCU
1	0	0	0
1	1	1	1
0	X	3-STATE	PIN

Figure 3-50: Typical port I/O circuitry used in the MC6805 family of microcomputers.

All pins on the MC6805P1 (ports A, B, and C, see Figures 3-44 and 3-45) are programmable as either inputs or outputs under software control of the corresponding Data Direction Register (see Figure 3-50). In this case, the user would program the corresponding bit in the Port Data Direction Register (DDR) to a logic 1 for output, or a logic 0 for input. On reset, all DDRs are initialized (to set addresses to their starting value or some prescribed points in the routine) to a logic 0 state to put the ports in the input modes. On the other hand, the port output registers are not initialized on reset but may have signals applied before setting the DDR bits, to avoid bits that do not fall into the active high or active low logic levels. *Example:* Logic high could be a minimum 3.5 volts, logic low, a maximum 0.4 volts, and other levels may be referred to an "undefined" level. To put it another way, the circuit will not respond at these levels.

When programmed as outputs, the latched output data is readable as input data, regardless of the logic levels at the output pin, due to output loading (see Figure 3-50). In case you are not familiar with *latch,* it is a term used to indicate data storage such as in a flip-flop. In fact, very often a D-type flip-flop is called a *latch* or *latching* circuit. Or we could say that a certain buffer has a latch function that permits each bit in the buffer to be latched to 1 or 0 by the data pulses. In this case, the latched byte (stored 8-bits) is held in the buffer until the register is ready to accept new data.

There are always cautions when servicing I/O ports on any MCU. For instance, the corresponding DDRs for ports A, B, and C (see Figure 3-50) are write-only (binary bit inputs) registers (registers $004, $005, and $006 in the MCU M6805P2). You cannot read these registers; that is, a read operation on these registers is undefined. Also, the latched output data bit (see Figure 3-50) may always be written. What this all boils down to is that any write to a port writes all of its data bits, even though the port DDR is set to input. You can use this to initialize the data registers and avoid undefined outputs. Nevertheless, you must take care when testing the read/modify/write instructions on the MC6805 family of MCUs, because the data read corresponds to the pin level of the DDR (is an output at 0), and corresponds to the latched output when the DDR is an output (1).

When working with microprocessors, etc., it is important you realize that data bytes can be transmitted in both parallel and serial form from the peripheral to the MPU. The parallel method is the fastest and requires the simplest I/O circuit in the MPU. But when using some peripherals, a *serial data bit stream* is the conventional way to transmit data. Figure 3-51 shows the relationship between a peripheral typewriter (using a digital in/out signal) and an IC called *Asynchronous Communications Interface Adapter* (ACIA). The ACIA permits data to be transmitted in serial format with only one line instead of the eight required for parallel transmission. An IC such as the MC6805 (an ACIA chip) can be used to convert from serial to parallel or vice versa. *Note:* MCUs such as the MC68701 have on-chip serial communications interface and parallel I/O.

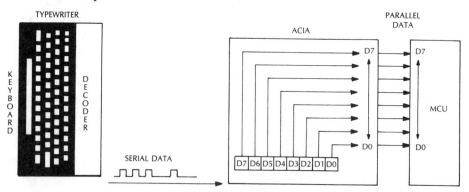

Figure 3-51: Input relationship between a digital serial output peripheral and a parallel input MCU, using an ACIA chip such as the MC6805, etc.

A considerable amount of today's computer equipment is being connected to telephone lines. In this case, you will sometimes find that another IC is required. This IC—for example, the MC6860, a modulator/demodulator (Modem)—is placed in series between the ACIA and telephone input/output line. A block diagram of such a hookup is shown in Figure 3-52.

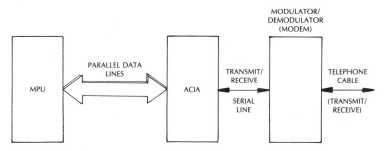

Figure 3-52: Modulator/demodulator, ACIA, and MPU. The input/output signal to the Modem from the telephone cable is a sinusoidal signal. All other lines, both serial and parallel, carry data in digital form. The purpose of the Modem is to make the conversion between sinusoidal and digital signals.

Motorola's extensive line of M6800 MPU peripherals interfacing ICs are directly compatible with their newer MC68000 family (16-bit MPU's). A few of these devices that you may encounter during servicing are listed in Table 3-11. Again, MCUs such as the MC68701 have both serial and parallel interface and, incidentally, this MCU also features on-board EPROM rather than maskable ROM.

PRODUCT NUMBER	NAME
MC6821	PERIPHERAL INTERFACE ADAPTER
MC6840	PROGRAMMABLE TIMER MODULE
MC6843	FLOPPY DISK CONTROLLER
MC6845	CRT CONTROLLER
MC6850	ASYNCHRONOUS COMMUNICATION INTERFACE ADAPTER
MC6852	SYNCHRONOUS SERAL DATA ADAPTER SERIAL
MC6854	ADVANCED DATA LINK CONTROLLER
MC68488	GENERAL PURPOSE INTERFACE ADAPTER

Table 3-11: Adapter and controller ICs for peripheral interface to M6800 (8-bit) and MC68000 (16-bit) MPUs.

Adapter and Control ICs (Registers)

Adapter and control IC's such as the ACIA MC6850 (see Table 3-11) contain various registers, just as do the MCU units. The MC6805 has four 8-bit registers that may be addressed:

1. Status Register (SR).

2. Receive Data Register (RDR). These are read-only registers (the MPU cannot write into them).

3. Transmit Data Register (TDR).

4. Control Register (CR). These last two registers (items 3 and 4) are addressable but not readable, i.e., the MPU cannot read data out of them, although it can instruct them.

In addition to these registers, there are control lines. Some of these are 3-chip select lines (CS0, CS1, CS2), one Register Select line (RS), one Interrupt Request line (IRQ), one Enable line (E), one Read/Write line (R/W), and seven Data Control lines. Many of these lines (pins 2, 5, 6, 23, and 24) are used to send and receive data when connected to peripheral equipment, or when the system includes a modem such as the MC6860 listed in Table 3-11. Figure 3-53 shows the pin configuration for the MC6850 ACIA.

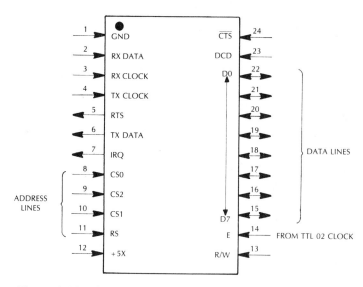

Figure 3-53: Pin configuration for the MC6850, an ACIA unit.

The Peripheral Interface Adapter (PIA), product No. MC6821, also has several registers: two peripheral data registers (A and B), two data direction registers (A and B), and two control registers (A and B). It also has two separate 8-bit bidirectional peripheral data buses that can be used for interfacing with external equipment. See Figure 3-54.

You will notice that the software features for these ICs (the MC6805 and MC6821) are similar to the port I/Os we discussed in the preceding section titled "Input/Output Signals," where we used the MC6805 MCU family as our example. Basically, the only difference is that the I/O circuitry is located on the MCU chip, but usually not on MPU chips. In other words, in most cases

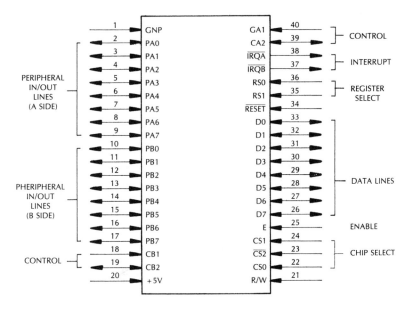

Figure 3-54: Pin configuration for the MC6821 listed in Table 3-11.

you would not need an ACIA with an MCU but would when connecting to a MPU.

The MCU has the necessary registers manufactured into the chip. There are 20 input/output pins on the microcomputer chip MC6805P2. All I/O pins (ports A, B, and C, see Figure 3-44) are connected to ports A, B, and C register. See Figure 3-50, which shows a typical port I/O circuit.

Timer Circuitry and How It Works

The electrical pulses used by the MPU, ROM, and RAM (whether they be internal in a MCU, or external when using a separate MPU) are generated by a *clock* or *timer*. It should be pointed out that these components, ROMs, RAMs, MPUs, etc., do not actually generate the binary pulses we have been discussing in previous pages, but produce their bytes on the address and data lines, when they receive clock pulses. Also, as we know, all digital systems must have a clock signal. Microcomputers such as the MC6805 family are no different. For example, the on-chip clock generator found on the MC6805 units uses an 8-bit timer, where others such as the MC68701 MCU feature a 16-bit 3-function *programmable* timer.

Inside an MCU such as the MC68701 is a timer (see Figure 3-55) that can be used to perform input waveform measurements while independently generating an output waveform. Pulse widths in this case can vary from several microseconds to many seconds.

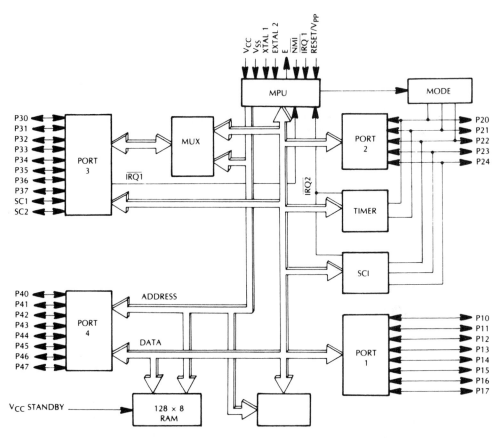

Figure 3-55: Block diagram for the MC68701 MCU. This chip is contained in a 40-pin package that uses ultraviolet erasure. Figure 2-69 (Section 2.13) shows a similar type of package with an ultraviolet window.

What makes this MCU timer special is that it can be programmed, as we mentioned. The key time element in this system is a 16-bit free-running counter. See Figure 3-56 for a block diagram of the timer. You will also notice that there are two 16-bit registers, the Output Compare Register and the Input Capture Register. The other register, the Timer Control and Status Register (TSCR), is an 8-bit register.

To illustrate the role of each block in the system, let us first see what they are and what they do. The mnemonics included with each heading apply only to the MC68701 family of MPU units. When a dollar sign is shown, it indicates the number is in hex.

Counter ($09:0A) The heart of the timer is the free-running 16-bit counter. This counter is incremented (caused to change state) by E (Enable, See Figure 3-55). That happens to be pin 40 on this chip. Timer Overflow

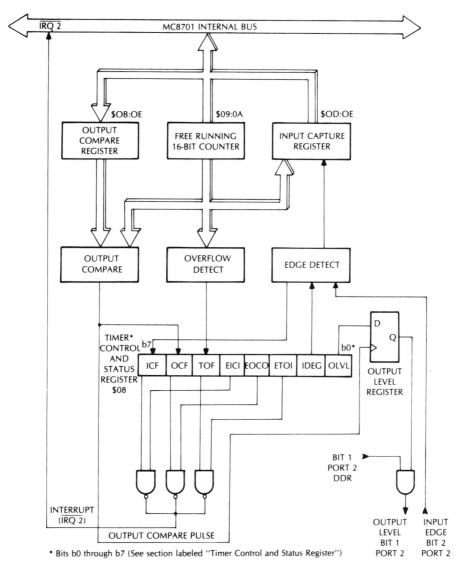

Figure 3-56: Block diagram of an on-chip programmable timer (MC68701 family).

Flag (TOF, see small center blocks in Timer Control and Status Register, Figure 3-56), is set whenever the counter contains all 1s. Let's digress for a moment and say a few words about "flag." Generally, a counter will send a *flag*, or signal, to other circuits, indicating a full count. This flag can be used for any number of functions. For instance, the flag can be used to halt or reset operation of the MPU, or it might be used to pulse another counter, etc.

Output Compare Register ($0B:0C) This register is a read/write circuit used to control an output signal waveform, or provide a timeout flag, based on one's preference. It is compared with the free-running counter (shown in Figure 3-55) on each cycle that is an input to the MPU in the same illustration. When a match occurs, Output Compare Flag (see OCF block in Figure 3-56) is set and the Output Level (OLVL) is clocked to an output control register that is shown in the block diagram. If port 2 (see Figures 3-55 and 56), bit 1, is configured as an output, OLVL will appear at P21 (see Figure 3-55) and Output Compare Register, and OLVL can then be changed for the next compare.

Input Compare Register ($OD:OE) This register is a read-only circuit used to store the free-running counter when a correct input signal change (transition) occurs. The IEDG presents the proper input for this operation (see block diagram, Figure 3-56). Port 2, bit 0, should be configured as an input; however, the Edge Detect circuit (shown just below the Input Compare

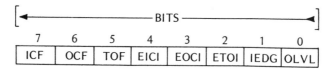

Figure 3-57: Digital bits b0 through b7 in the Timer Control and Status Register. See Table 3-12 for function description.

TCSR 8-BITS	WHAT EACH BIT DOES
BIT 0 OLVL	Output level. OLVL is clocked to the output level register by a successful output compare and will appear at P21 if Bit 1 of the Port 2 Data Direction Register is set. It is cleared during reset.
BIT IEDG	Input Edge. IEDG is cleared during reset and controls which level transition will trigger a counter transfer to the Input Capture Register.
BIT 2 ETOI	Enable Timer Overflow Interrupt. When set, an IRQ2 interrupt is enabled for a timer overflow; when clear, the interrupt is inhibited. It is cleared during reset.
BIT 3 EOCI	Enable Output Compare Interrupt. When set, an IRQ2 interrupt is enabled for an output compare; when clear, the interrupt is inhibited. It is cleared during reset.
BIT 4 EICI	Enable Output Capture Interrupt. When set, an IRQ2 interrupt is enabled for an input capture; when clear, the interrupt is inhibited. It is cleared during reset.
BIT 5 TOF	Timer Overflow Flag. TOF is set when the counter contains all 1's. It is cleared by reading the TCSR (with TOF set), then reading the counter high byte ($09), or by RESET.
BIT 6 OCF	Output Compare Flag. OCF is set when the Output Compare Register matches the free-running counter. It is cleared by reading the TCSR (with OCF set) and then writing the to the Output Compare Register($0B or #0C), or by RESET.
BIT 7 ICF	Input Capture Flag. ICF is set to indicate a proper level transition; it is cleared by reading the TCSR (with ICF set) and then the Input Capture Register High Byte (%0D), or by RESET.

Table 3-12: Timer control and status register event table for the programmable timer manufactured in the MC68701 microcomputer chip.

Register in the block diagram) always senses P20 (see Figure 3-55), even when it is used as an output. But an input capture can occur independently of ICF (shown in the Timer Control and Status Register): the register always contains the most current value. Also, the input pulse width must be at least two E cycles, if it is desired to capture under all conditions.

Timer Control and Status Register ($08) We have repeatedly referred to this register, therefore let's examine its operation in more detail. First, refer to Figure 3-57, which is an extract from the center of the block diagram shown in Figure 3-56, and which shows a digital signal (bits b0 through b7).

This register, the Timer Control and Status Register (TCSR), is an 8-bit device with all bits readable, while bits 0 ... 4 can be written. The three most important bits provide the timer status and indicate the following factors:

1. If a correct bit level transition has been detected.
2. Whether a match has occurred between the free-running counter and the output compare register.
3. Whether the free-running counter is in an overflow conditon.

Any one of these conditions can generate an IRQ2 (interrupt) signal and is controlled by an individual enable bit in the TCSR (see block diagram in Figure 3-56). Table 3-12 lists each function, starting with bit 0 (OLVL) and ending with bit 7 (ICF), shown in Figures 3-56 and 57.

Addressing Modes

In Section 3.5, under the heading "How and Why Addressing Modes Are Used in a Microcomputer System," we discussed the fundamental addressing modes for the M6800 family. As you might expect, the more advanced model, the MC68701 MCU, is very similar. It is an 8-bit, single-chip MCU unit that has 6 addressing modes that can be used to reference memory:

1. *Inherent* (the operand): Registers, and no memory reference is required. These are 8-bit instructions.
2. *Immediate:* As explained in Section 3.5, the operand or immediate byte(s) of the instruction where the number of bytes matches the size of the register. Also, these are 2 and 3-byte instructions.
3. *Direct:* This is a 2-byte instruction. The least significant byte of the operand address is contained in the second byte of the instruction and the most significant byte is assumed to be $00.
4. *Extended:* As explained previously, in this addressing mode the next instruction to be executed by the MPU is located at some other address than the one following.
5. *Indexed:* The indexed mode of addressing uses a 2-byte instruction. The number contained in the second byte (next location after the

instruction) is often referred to as an "offset." The unsigned offset contained in the second byte of the instruction is added with Carry to the index register and used to reference memory without changing the index register. This new address, as in the extended mode, contains the data.

6. *Relative:* As explained in Section 3.5, in this addressing mode the next instruction to be executed by the MPU is located at some address other than the one following.

In other words, this mode is used only for branch instructions. These are 2-byte instructions and, if the branch condition is true, the program counter is overwritten with the sum of a signal single byte displacement in the second byte of the instructions and current program counter. What this does is to add many addresses—branch range; for example, relative addressing mode for the MC68701 MCU provides a branch range of (126 to 129) bytes from the first byte of the instruction.

Valid Machine Codes

Although a single instruction can have several modes, it is important to realize that each addressing mode (Inherent, Immediate, Direct, etc.) is just a different way of telling the MCU where to reference memory. For example, the Motorola data sheet for their MC6800 lists 72 instructions—variable length; but, as we know, each instruction can have more than one addressing mode. In the case of the MC6800 MPU, it adds up to 197 valid machine codes. Newer units such as the MC68701 MCU have more—82 instructions—and, in this case, there are 220 valid machine codes. Table 3-13 lists seven of the possible accumulator and memory instructions for the MC68701 MCU.

Don't forget, the mnemonic (MCE) code is assigned by the manufac-

ACCUMULATOR AND MEMORY OPERATIONS	MNE	IMMED			DIRECT			INDEX			EXTEND			INHER		
		OP	~	#	OP	~	#	OP	~	#	OP	~	#	OP	~	#
ADD ACMLTRS	ABA													1B	2	1
ADD B TO X	ABX													3A	3	1
ADD WITH CARRY	ADCA	89	2	2	99	3	2	A9	4	2	B9	4	3			
	ADCB	C9	2	2	D9	3	2	E9	4	2	F9	43	3			
ADD	ADDA	8B	2	2	9B	3	2	AB	4	2	BB	4	3			
	ADDB	CB	2	2	DB	3	2	EB	4	2	FB	4	3			
ADD DOUBLE	ADDD	C3	4	3	D3	5	2	E3	6	2	F3	6	3			
AND	ANDA	84	2	2	94	3	2	A4	4	2	B4	4	3			
	ANDB	C4	2	2	D4	3	2	E4	4	2	F4	4	3			
SHIFT LEFT ARITHMETIC	ASL							68	6	2	78	6	3			
	ASLA													48	2	1
	ASLB													58	2	1

Table 3-13: Seven example accumulator and memory instructions for the MC68701 MCU. This chip has a total of 82 separate instructions and, because of separate addressing modes, has 220 valid machine codes.

turer. For instance, ABA (Add accumulator B to A) is represented by the hex code 1B (see column titled Inherent Mode, under Operation Code OP). This would actually appear in ROM memory as 00011011 (see Table 2-2; the extra three 0s are needed in an 8-bit system). Also, the symbol shown in Table 3-13 stands for the number of cycles and the other symbol, #, is read as the number of program bytes.

In the previous pages it was said that when a dollar sign is shown, it indicates the given number is in hex. However, there are other symbols that are used to indicate certain functions, etc. For instance, when a # sign is shown, the instruction is in the immediate mode, and the number following this sign is located in the next byte of memory. As another example, the index addressing mode may be shown with a hex number, say 19 (00011001 binary, 25 decimal), followed by a comma and X, i.e., it could be $19,X. In this case, the number following the $ is added to the contents of the index register to form a new effectual address.

Introduction to Programming Basics

Although you should now know that a program for a computer or processor consists of a sequence of operational instructions stored in memory, it is important for you, as a user, to know the MPU, MCU language, and the manner in which the device operates. In short, you must know the *Instruction Set*. The set of all instructions common to a given MPU, MCU, or CPU is referred to as its Instruction Set. Each instruction in the Instruction Set (the writing of the Instruction Set is the task of the programmer) enables a single elementary operation such as the movement of a data byte, an arithmetic or logical operation on a data byte, or a change in instruction execution sequence.

A point that bears repeating, and one you may have missed, is that the size of the Instruction Set is a measure of the capabilities of the MPUs, the MCUs, or the CPUs. Another such measure you may have noticed is the length of binary words the device can work with (usually 4-bit, 8-bit, or the new 16- and 32-bit). Generally speaking, the larger the Instruction Set or word size, the more powerful the MPU, MCU, etc. The MC68000 family of MPUs, 16-bit MPU with over 1000 useful instructions (this includes the Instruction Set and all variations of instructions) is thus more powerful than the MC141000, a 4-bit MCU with 43 standard instructions.

After the programmer has written the program, it is stored in memory (EPROM, ROM, or RAM) as a sequence of bytes that represent the instruction. Then to review a bit, the program execution proceeds sequentially (i.e., memory location 8001 is executed after location 8000, etc.), unless a transfer-of-control, or branch instruction (branch-forward, branch-back, or jump, for example) is executed, which causes the PCU to set to a specified memory address. *Note:* The instruction *jump* causes the MPU to branch to another program (usually called a *subroutine*—a program within a program). Sets of rules or processes for solving a problem in a certain number of steps— example, arithmetic procedure—are often programmed as subroutines.

The subroutine type of jump requires the MPU to store the contents of the program counter at the time the jump occurs. This enables the processor to resume execution of the main program after the last instruction of the subroutine has been executed (called *Return from Subroutine,* or RTS).

The *Stack Pointer,* a 2-byte register that was mentioned in Section 3.5, contains a beginning address, normally in RAM, where the status of the MPU's register may be stored during a branch to a subroutine. A subroutine may call up another subroutine. This is called *nesting subroutines.* If the MCU, etc. being used has a stack for storing return addresses, the maximum depth of nesting subroutines is determined solely by the depth of the stack itself.

The MC8705R3, an 8-bit EPROM MCU with analog-to-digital conversion, is capable of subroutines and interrupts that may be nested down to an on-chip location $061 (31 bytes maximum), which allows the programmer to use up to 15 levels of subroutine calls (fewer, if interrupts are allowed). *Note:* Interrupts, as you would imagine, refer to a method in which the MCU is interrupted from doing its primary duties so that it may perform a certain more important task.

SECTION 3.7: DARLINGTON TRANSISTOR ICs

Driving peripherals often requires more current and/or voltage than some ICs can deliver. There are Darlington amplifier transistor ICs that are

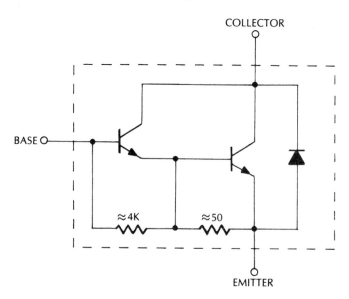

Figure 3-58: NPN silicon power Darlington transistor (120 W)

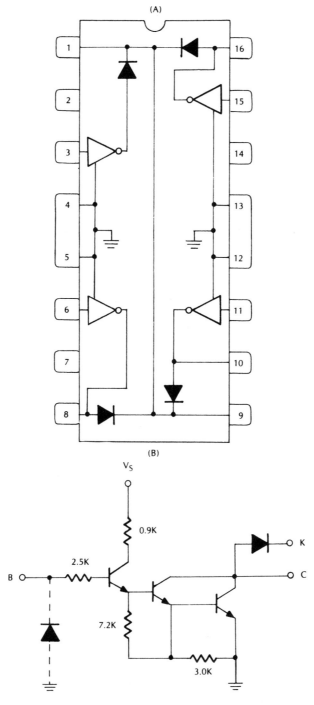

Figure 3-59: Octal Darlington array. ULN2803, with output clamp diodes.

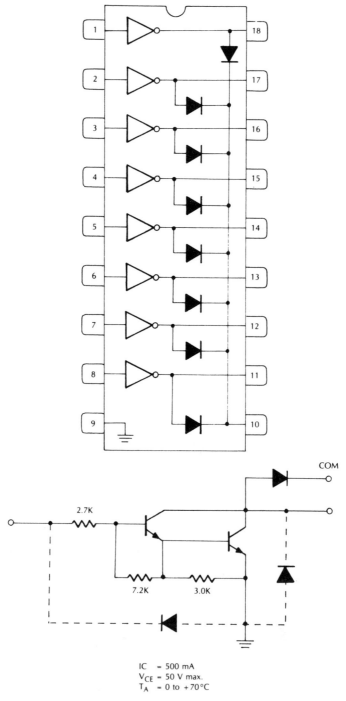

IC   = 500 mA
V_{CE} = 50 V max.
T_A = 0 to +70°C

Figure 3-60: Quad driver array using Darlington circuit designed to utilize three transistors.

specifically designed for amplifier and driver applications where high gain is essential. As an example, there are the NPN D40C1, 2, 3, and 5 Darlingtons. The D40C2 and 5 produce a minimum current gain of 40k at an I_c of 200 mA, with V_{CC} = 5 Vdc. *Note:* DC current gain = β = h_{fe} = I_c/I_b at a certain ambient temperature (usually 25°C). Incidentally, the maximum h_{fe} for all four of these ICs is 60 k, with an I_c of 200 mA at 5 Vdc. Another example of a two-transistor Darlington IC, the 2N656, 7, and 8 series, is shown in Figure 3-58.

Next, the input impedance. In the common collector mode of operation, the input impedance is approximately equal to $h_{fe}^2 \times R_L$. The reason for the high current gain of Darlingtons is the fact that the h_{fe} is approximately equal to the average h_{fe} of the two transistors, squared.

Darlington ICs are not limited to a single Darlington amplifier. Several Darlingtons can be used in one package. A state-of-the-art example of this is Motorola's peripheral interface driver array ULN2803, an "octal Darlington array." See Figure 3-59 for pin connections and a single element Darlington schematic for this IC. Furthermore, Darlington circuits need not be limited to two transistors. Three (and even four) transistors can be used in the Darlington circuit. An example of this is Motorola's circuit in Figure 3-60 (B). This IC is a quad driver array (ULN2968), 1.5 V, V_{CE} = 50 V max (see Figure 3-60 (A) for logic diagram). The circuit is essentially a common emitter Darlington followed by a common collector Darlington.

SECTION 3.8: POWER MOSFETs

In the past few years, several semiconductor manufacturers have brought out a new generation of power field-effect transistors that are invading the field of bipolar power transistors.

These power transistors are known by several names (depending on the manufacturer and manufacturing technique): Motorola's TMOS power MOSFET, Siemen's SIPMOS, and International Rectifier's HEXFET, plus others. But, whatever they are labeled, in general they all rely upon vertical current conduction through the transistor. Figure 3-61 is a pictorial view of Motorola's TMOS power FET. Note the label *drain current* and arrows pointing out the direction of drain current flow. These four arrows (two coming from the right side, two coming from the left side, and then down) seem to form the letter T; whence the name TMOS. See Figure 3-62 for schematic and package types.

TMOS and Bipolar Power Transistors: Advantages and Disadvantages

It's important to realize that, in some cases, you are better off using bipolar power transistors rather than MOS power transistors. Here are some of the advantages and disadvantages of these power TMOS FETs:

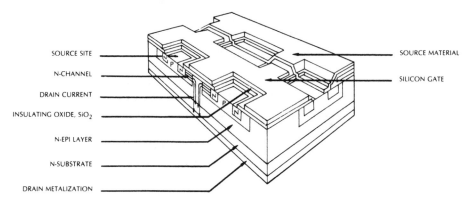

SOURCE SITE

N-CHANNEL

DRAIN CURRENT

INSULATING OXIDE, SiO$_2$

N-EPI LAYER

N-SUBSTRATE

DRAIN METALIZATION

SOURCE MATERIAL

SILICON GATE

Figure 3-61: Pictorial view of Motorola's TMOS power field-effect transistor. Note vertical current flow in FET structure. This offers a low resistance path and permits smaller chip size, resulting in a major breakthrough in the MOS power transistor. See Figure 3-62 for actual case and schematic for these devices.

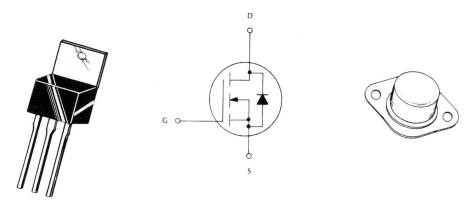

Figure 3-62: Schematic and package types for Motorola's MTM, MTP, N-channel TMOS power FETs.

1. The FETs have high-impedance input that could provide a better impedance match for the driving signal source you are interfacing to.
2. Bipolar power transistors are a current-driven device. FETs are voltage driven. It's possible to use simpler drive circuits in some applications using a TMOS power transistor. To put it another way, TMOS FETs are easy to drive and interface because of their low voltage drive requirements.
3. MOS power transistors have significantly faster switching speeds than bipolar. Nanosecond speeds are possible with the TMOS transistors.

4. The forward conduction ("on") resistance may be as small as a few tenths of an ohm. Typically, it's from about 0.25 to 2 ohms for the IRF, MTM, and MTP series.

5. TMOS power transistors are more stable than bipolar transistors. This is because as chip temperature rises, current drain decreases.

6. The principle disadvantage of FETs, in respect to bipolars, is that they are susceptible to electrostatic discharge. The TMOS FET is no different. For this reason it also requires special handling.

SECTION 3.9: WORKING WITH THYRISTORS

Thyristors are basically a bistable device comprising three or more junctions. In general, they consist of two generic component categories—silicon-controlled rectifiers (SCRs), and triacs (a bidirectional rectifier, essentially two SCRs in parallel). Some of these switches (triacs) are designed to be triggered directly by microprocessors and microcontrollers. Motorola's M4C228-2 - 10 family is well suited for interfacing personal computers with line-operated appliances and similar devices. Figure 3-63 shows seven schematic symbols representing typical thyristors used in today's electronic industry. See Section 10 for thyristor testing procedures.

Unijunction Transistors (UJT) Having briefly described SCRs and triacs, let's look at the others. The UJT (Figure 3-63) is a unijunction transistor. These are usually a highly stable IC used for general purpose *trigger applications* and, as a general rule, not expensive. You'll also find these transistors in pulse-generating (oscillators) and timing circuits.

Bilateral Triggers (DIAC) The next symbol shown (Figure 3-63) is known as a DIAC. Although the symbol looks like an incomplete transistor (because there is no base lead, and there appear to be two emitters), it is not. In reality, the device is a bidirectional breakdown diode that conducts only when a certain breakdown voltage is exceeded. For example, a 1N5758A DIAC (also called a *bilateral trigger*) has a switching voltage range of 16 to 24 volts (both directions). The switching current (both directions) is 100 μA, and the switching voltage change (both directions) is 5 volts. One application where you might find this DIAC could be the control circuit for a triac.

Programmable UJTs The PUT symbol (see Figure 3-63 E), depicts a *programmable unijunction transistor.* These transistors physically look like a UJT (which is the same as almost any bipolar transistor, i.e., three leads out of a metal or plastic case) and are similar to UJTs, except that they can be programmed with external "program" resistors—usually labeled R_1 and R_2. Figure 3-64 shows a PUT with "program" resistors R_1 and R_2.

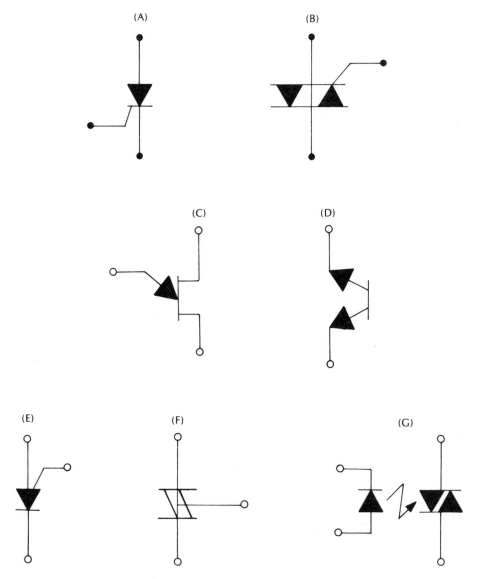

Figure 3-63: Schematic symbols of modern thyristors.

When calculating the value of the gate voltage (V_S) and determining the value of V_B) shown in the formula for V_S in Figure 3-64, it is important to refer to the maximum ratings of the PUT. This example is for the 2N627 and 2N628 programmable unijunction transistors (40 volts, 375 mW). The low

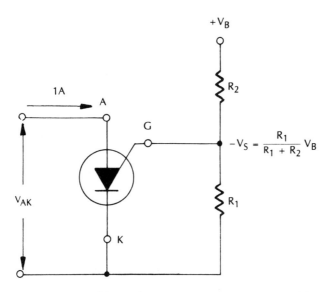

$$+V_B$$

$$R_2$$

$$-V_S = \frac{R_1}{R_1 + R_2} V_B$$

$$R_1$$

Figure 3-64: Programmable unijunction transistor (PUT) with "program" resistors R_1 and R_2.

MAXIMUM RATINGS			
RATING	SYMBOL	VALUE	UNIT
Power Dissipation	P_f	300	mW
Derate above 25°C	$1/\theta_{JA}$	4.0	mW/°C
DC Forward Anode Current	I_T	150	mA
Derate Above 25ξ 25°C		2.67	mA/°C
D DC Gate Current	I_G	± 50	mA
Repetitive Peak Forward Current			
100µs Pulse Width, 1.0% Duty Cycle	I_{TRM}	1.0	Amp
20µs Pulse Width, 1.0% Duty Cycle		2.0	Amp
Non-Repetitive Peak Forward Current			
10µs Pulse Width	I_{TSM}	5.0	Amp
Gate Gate to Cathode Forward Voltage	V_{CKF}	40	Volt
Gate to Cathode Reverse Voltage	V_{GKR}	− 5.0	Volt
Gate to Anode Reverse Voltage	V_{GAR}	40	Volt
Anode to Gate Voltage	V_{AK}	± 40	Volt

Table 3-14: Maximum ratings for the PUT 2N6027/8

"on" state for these devices is 1.5 volts maximum at a forward conducting current (I_F) of 50 mA. Some useful maximum ratings for these transistors are listed in Table 3-14.

Silicon Bidirectional Switches (SBS) Next, let's look at the SBS symbol (Figure 3-63 F). The letters SBS are an abbreviation for Silicon Bidirectional

Switch. The application for this solid state device is similar to the DIAC we have explained, but it has gate electrodes that permit synchronization. Two types are MBS4991 and MBS4992. The main differences between the two are switching voltage and current ratings, plus other electrical characteristics such as forward on-state voltage, holding current, etc.

These bidirectional switches are also called *Bidirectional Diode Thyristors* and they are designed for full-wave triggering in triac phase control circuits, half-wave SCR triggering applications, and voltage level detectors.

Optically Coupled Drivers Figure 3-63 G, the symbol for an optically coupled triac driver, contains an *infrared* LED and a bidirectional photodetector in one isolated plastic DIP. These optically isolated devices are used as a triac driver and, typically, have a high isolation voltage (I_{SO}) rating: $V_{ISO} = 7500$ minimum.

Motorola's MOC3030 and MOC3031 are six-pin ICs that are designed for use with a triac in the interface of logic systems, to equipment powered from 120 Vac power lines. There actually are many applications for these devices and all related ICs (there are quite a few on the market today). For instance, computer viewing screens (CRTs), printers (hard copy machines), and almost any similar application where protection of expensive equipment and/or personnel is desirable, could use an optical isolater.

SECTION 3.10: OPTOISOLATORS, PHOTOTRANSISTORS, AND PHOTO DARLINGTON AMPLIFIER ICs

Optoisolators

In the typical LED/phototransistor optoisolator, a transparent glass window called a *dielectric channel* isolates and insulates the LED chip from the phototransistor. By this means, an isolation of 25,000-75,000, etc., volts is readily achieved. There are quite a few different source/detector combinations. ICs such as the MOC3030 can be shown schematically. See Figure 3-65.

As was explained in Section 3.9, one of the most important applications for optoisolators is high-voltage isolation for expensive equipment such as computers and for personnel safety. Figure 3-66 is an illustration of how you might use a MOC3030 in a hot line switching application circuit. See Section 10 for optoisolator test procedure.

In this circuit, Motorola has shown the load connected to the neutral side of the 120 Vac line. However, you can place the load in either side if you remember that the load will be "hot" in reference to neutral (usually ground) if you place it in the hot side of the 120 Vac line. *Note:* Although this can be done, it is *not* recommended!

The value of R_{in} shown connected to pin 1 (the anode of the internal

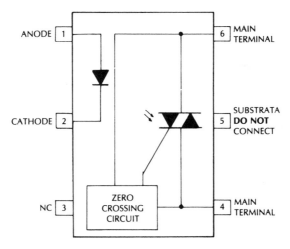

Figure 3-65: Schematic and pin diagram for a MOC3030 and MOC3031 opto-coupler, zero-crossing triac driver.

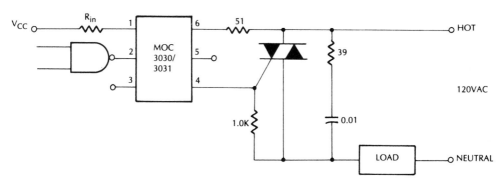

Figure 3-66: Typical circuit using a MOC3030/31 when hot line switching is required.

diode—see Figure 3-66) can be calculated if you know the continuous forward current rating of the IC port. The forward current (I_F) is equal to the rated trigger current (LED trigger current, I_{FT}), which is 30 mA for the MOC3030, and 15 mA for the MOC3031. Also, the resistor and capacitor connected in series (the 39 ohm resistor and 0.01 μF capacitor) may not be needed with some triacs and loads.

Using Opto Coupler/Isolators for TTL to MOS Interface

The 4N35, 6, and 7 opto coupler/isolator with transistor output that is constructed with a NPN phototransistor and PN infrared emitting diode may

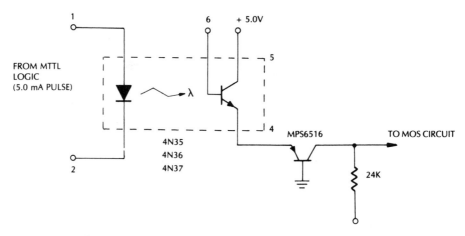

Figure 3-67: Using an opto coupler/isolator to interface TTL to MOS (P-channel).

be used as a TTL to MOS (P-channel) interface. Figure 3-67 shows how this could be done.

Photo Darlington Amplifiers

Figure 3-68 shows the internal schematic of a photo Darlington amplifier with an output circuit (R_L) included. These plastic NPN silicon photo

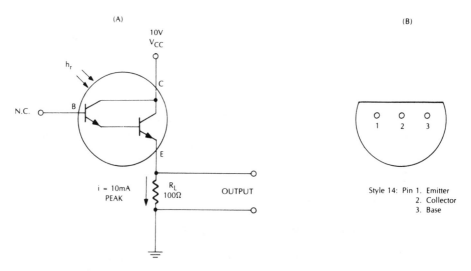

Figure 3-68: Photo Darlington amplifier 2N5777 through 1N5780 (12, 25, and 40 volt). (A) Schematic diagram. (B) Package with pin identification.

Darlington amplifiers have a clear plastic case that permits visible and near-infrared light to impinge on the silicon photo detector. They have extremely high radiation sensitivity and are very stable. A less expensive device that will do basically the same job is the phototransistor or photodiode. However, phototransistors are only moderately sensitive compared to the photo Darlingtons. The photodiodes' claim to fame is their high speed (for example 1.0 ns).

SECTION 3.11: PRACTICAL MOUNTING GUIDE
FOR POWER SEMICONDUCTORS

It is important you realize that for current and power ratings of a power transistor, thyristors, or any other semiconductor devices such as low current, lead mounted transistors, MOS ICs, small signal and other devices not requiring heat sink mounting, some form of heat dissipator is required to prevent the component's internal temperature from rising above the manufacturer's rated value. If a heat sink and/or fan is not used, you are running the risk of destroying a power semiconductor device.

Poor mounting practices can result in failures of power semiconductors. This is particularly true when any of the various plastic packaged semiconductors are used. For instance, bending the leads to fit a socket has caused the untimely death of many an electronic component. Figure 3-69 shows the mounting hardware for a T0-126 packaged semiconductor. These packages are called *Thermopad©* plastic power packages by Motorola. An important feature of mounting this type of package is the compression washer shown.

The life span of semiconductors packaged in this manner depends largely upon whether you use a compression washer, and how well you burr the hole you drill into the mounting surface. *The heat sink cannot be perforated board.* It *must* be a *metal base.* Also, do not tighten the machine screw any more than is needed to apply the proper force to the bell compression washer (*don't* flatten the washer excessively).

You will find that machine screws and nuts are very good fasteners for all types of semiconductor packages that have a hole in them. But, again, use a compression washer! You can use sheet metal screws in some cases, but be careful that they do not lift the semiconductor package off your heat sink. To improve contact between the package and heat sink, it is frequently better to use one of the thermal greases on the market. *Note:* In some cases, manufacturers recommend that you do not use the thermal compounds when mounting a certain device. This information is included in the semiconductor spec sheets.

Specific data for each class of package is given in separate data sheets for each package, and manufacturers usually supply mounting hardware for all power packages. Nevertheless, as a general rule you can do your own

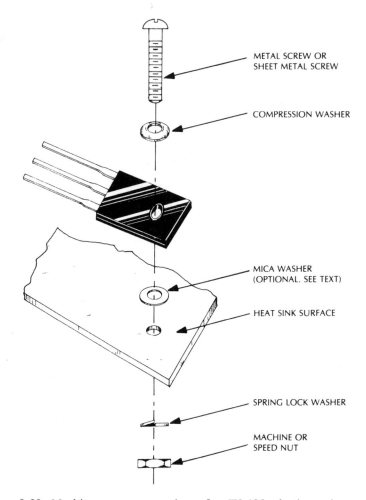

METAL SCREW OR
SHEET METAL SCREW

COMPRESSION WASHER

MICA WASHER
(OPTIONAL. SEE TEXT)

HEAT SINK SURFACE

SPRING LOCK WASHER

MACHINE OR
SPEED NUT

Figure 3-69: Machine screw mounting of a T0-126 plastic package power semiconductor.

package mounting of medium-power semiconductor devices if you are careful and follow the instructions on the manufacturer's spec sheet.

Guidelines for Electrical Insulation of Power Semiconductors

The problem of insulating the case (usually, the anode) of power semiconductors is a common one for many technicians. For the best results, isolate the entire assembly—semiconductor device and heat sink—from the chassis (ground). An example of this type of isolation is shown in Figure 3-70.

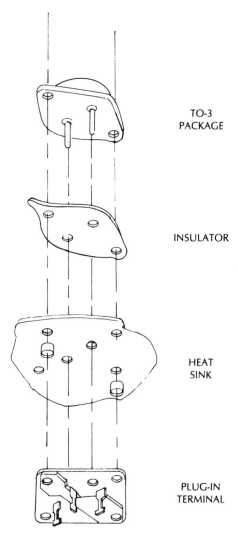

TO-3
PACKAGE

INSULATOR

HEAT
SINK

PLUG-IN
TERMINAL

Figure 3-70: Isolating a flat-mount power semiconductor from a chassis. Notice, the heat sink also is not connected to the chassis. The socket center plug-in terminal is the case (anode) connection.

There are times when you will find that it is not practical to isolate the entire assembly, as shown. In these cases, use an insulator between the chassis and semiconductor. But, in this type of mounting, in almost every case a thermal compound (for example, Radio Shack's "heat sink compound," a silicon-base grease) becomes increasingly important. When you are working with *high power*, it it best to specifically ask for an insulating grease manufactured for this purpose. What you want is a reduction of thermal resistance between the semiconductor and heat sink. Remember, clean your work surface before applying the heat sink compound and be sure your mounting screws are tight.

SECTION 4

Practical Radio Frequency Equipment Test and Measurement Guide

SECTION 4.1: BASIC PRINCIPLES OF RADIO FREQUENCY SOLID STATE SYSTEMS AND DEVICES

The fast-growing field of home satellite TV systems, cable TV systems and direct broadcast satellites, in both the United States and Canada, offers opportunities for the progressive service technician who understands radio frequency (rf) semiconductors. Figure 4-1 clearly illustrates that high frequency solid state semiconductor packages are different in appearance from the standard audio frequency transistor packages with which most of us are familiar.

This **section** not only includes an introduction to the subject of rf fundamentals needed by anyone dealing with rf equipment and today's semiconductors (like those shown in Figure 4-1); it also shows you what to do as well as how to do it.

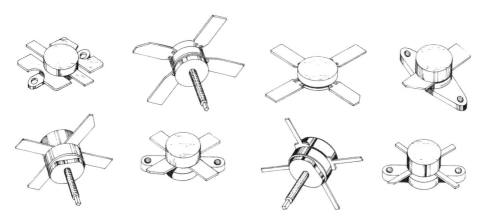

Figure 4-1: State-of-the-art packages used for rf power transistors.

Frequency Spectrum

The term *frequency* has different meanings for different electronic technicians. For example, a person working with audio systems will usually think of audio frequencies (abbreviated af) as falling roughly between 15 cycles per second (Hz) and 20,000 Hz. But someone working on rf equipment will probably think of a frequency range of something like 10 kHz through 30,000 MHz. Actually, frequencies are divided into many subgroups. Table 4-1 lists the frequency bands. Referring to this table, radio frequency ranges from VLF to EHF (3 kHz to 300 gHz).

FREQUENCY RANGE		DESIGNATION
30 to 300 hertz	ELF	Extermely Low Frequency
300 to 3000 hetz hertz	VF	Voice Frequency
3 to 30 kilohertz	VLF	Very-Low Frequency
30 to 300 kilohertz	LF	Low-Frequency
300 to 3000 kilohertz	MF	Medium-Frequency
3 to 30 megahertz	HF	High-Frequency
30 to 300 megahertz	VHF	Very-High Frequency
300 to 3000 megahertz	UHF	Ultra-High Frequency
3 to 30 gigahertz	SHF	Super-High Frequency
30 to 300 gigahertz	EHF	Extremely-High Frequency

Table 4-1: Frequency bands.

Photosensors: Frequency versus Wavelength

In Sections 3.9 and 3.10, we discussed photo transistors and other opto devices such as photo diodes, etc. (see Section 10 for testing procedure). It was pointed out that many of these semiconductors operate in the near-infrared band. A few years ago, infrared was often described in terms of Hertz. Today, however, you will usually find that infrared is considered in terms of wavelengths. Typically, these range all the way from 300 to 3 microns or micrometers in wavelength. Figure 4-2 shows the wavelength in micrometers (μm) for a pulsed infrared emitting diode, in reference to the constant spectral response.

Photosensors such as photo transistors respond to the entire visible radiation range as well as to infrared. The visible spectrum ranges in wavelength from 0.7 to 0.4 μm. Incidentally, it should be pointed out that all silicon photosensors (diodes, transistors, Darlingtons, triacs) show the same basic radiation response to the same wavelength; i.e., they respond to the near infrared peak, as shown in Figure 4-2.

As you can see, when dealing with electromagnetic waves it is often preferable to use wavelength (λ) rather than frequency. Generally, in situations that use wavelength, it is because wavelength is more easily measured than frequency. In such a case, frequency would be calculated from the measured wavelength using the familiar formula: frequency in Hz = speed of light/wavelength in meters.

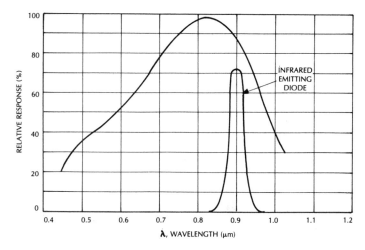

Figure 4-2: Infrared light-emitting diode spectral responses.

RF Semiconductor

When working with semiconductor devices that are designed to operate in the VHF part of the spectrum and in the so-called *microwave band*, you can easily identify them because they usually require special designs for the

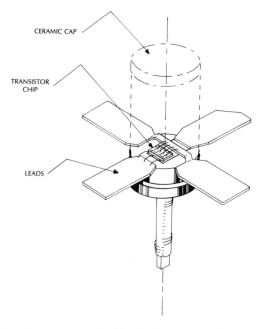

Figure 4-3: Low inductance strip-line leads connected to a transistor chip.

terminations (leads) of the transistors, diodes, etc. These designs (see Figure 4-1) are required to minimize the inductance caused by the pin-connecting arrangement.

The devices shown in Figure 4-1 are called *Strip-line Opposed Emitters* (SOE) transistor packages. Figure 4-3 shows how the transistor chip leads are connected to the flat rectangular strips used as connecting leads. This illustration is shown in a submounted package (an SOE package) with the top cap removed so that you can see the inside connections of the transistor.

Impedance (at rf frequencies)

If an impedance mismatch exists between the rf power transistor, the transmission line (for example, the strip-line leads shown in Figure 4-3) and the next stage (the load), some of the transmitted energy will be reflected back into the transmission line. This will set up what are called *standing waves* of voltage and current within the system and, most important, will possibly damage a high power component (and certainly cause a loss of energy).

This means that, for the best possible energy transfer from stage to stage, *impedance matching* is necessary when you are working with rf systems. For example, if the signal source impedance is 50 ohms (a typical value), the transmission line (coaxial cable, etc.) is 50 ohms, and the input impedance of the load is 50 ohms. The impedance of the system is matched and there will be maximum transfer of energy from the source to the load.

The output impedance of rf power transistors is usually given by all manufacturers, on the transistor data sheets, as a capacitance C_{out}. In general, the capacitive reactance output impedance will more nearly approach a resistive output impedance (which is what you want) in the upper part of the transistor operating bandwidth. However, the load that must be presented to an rf transistor is frequently indicated on the transistor data sheet.

Matching Networks

There are numerous basic impedance-matching networks that can be used when one must connect electronic components having different impedances into the same system. For instance, at sufficiently high frequencies (commercial TV frequencies and higher), matching networks using quarter-wave transformers may be used to accomplish impedance matching. Matching networks normally take the form of filters—both passive and active. As an example of an active filter, an OP AMP with capacitive feedback (voltage leads current by a given amount of degrees) can be designed to act like a circuit that will appear as an inductance, or even an entire LC network.

In summary, every rf system, no matter what its physical form, will have a definite value of impedance at the point where the transmission line ((coax, or whatever) is to be connected to each component (rf transistor amplifier, rf signal generator, or any other rf device). The problem is to transform this impedance to the proper value to match the connecting lines so that all components of the entire system are matched to each other *ohm for ohm*.

As a simple example, most of us know that a TV antenna has an output impedance of 300 ohms. Next, the input impedance to the TV receiver is 300 ohms. If each of the components is not the same impedance, it must be matched by using a matching network. For instance, if there is a 75-ohm coaxial cable coming down from a 75-ohm antenna, you would need to install a 75-to-300 ohm matching transformer to match the down lead to the receiver. Remember, impedance matching is important in any rf servicing job, whether it's connecting test gear, antennas, or rf transistors in a circuit.

Testing Solid State Radio Frequency Systems

When servicing, you have one slight technical advantage with very low-priced solid state rf systems, whether they are CB radios, FM two-way radios, aircraft transceivers, marine radio telephones, or ham gear. Their design is simple. For example, budget-priced CB radios' front panels contain only volume, squelch, and channel knobs. Generally there is an LED channel readout, and occasionally there's a meter, and sometimes a switch—but that's about it. Low-cost CB radios are simpler, by far, than their more expensive cousins: aircraft transceivers, marine radio telephones, and other sets such as deluxe AM/SSB mobile CB rigs.

When testing complete rf systems such as these, connect for transmitter testing first. Solid state transmitters are easier to service and quicker to diagnose than solid state receivers. Your first step is to measure the transmitter rf power (unmodulated). If you are an amateur radio operator working on a ham rig, etc., you need a test meter. One that you could use is the Power/Modulation/SWR meter sold by Radio Shack. Typically, low-cost meters such as these will measure power output in three ranges: 0 - 5 watts for CB, 0 - 5 watts and 0 - 500 watts for amateur band equipment. Frequently the same set measures percent of modulation from 1 to 120 percent, and can also be used to check the standing wave ratio (SWR) of the antenna system.

Another piece of test gear that one should have when checking rf systems is a field strength meter. This test instrument will help you operate the rf system with maximum efficiency. Generally, they are quite inexpensive, and they will measure SWR forward and reflected power to 1000 watts. Most of the CB types have a 52-ohm characteristic impedance and operate between 2 and 30 MHz.

It is important you realize that measuring output power, SWR, or percent of modulation does not require an FCC operator's license. Nor is an FCC license required to work on a transmitter if it is connected to a *nonradiating* dummy antenna. However, if any repairs or changes that affect power out, modulation, or frequency are made to a transmitter circuit, the system *must* be checked by a licensed technician before it is connected to a radiating antenna.

To check communications receivers, you will need the right signal generator and a good diagnostic ability. But ICs, in either a transmitter or a receiver, should not worry you much. For example, a certain input signal (usually given on the equipment schematic or spec sheet) should produce a specified output signal. If it doesn't, check the IC's dc voltages. If they are OK and you still don't get the proper output signal, the IC is probably defective and you should replace it.

Other than the special equipment listed, you will need the standard shop equipment (assuming you are starting a new shop). This includes a voltmeter and probes (both rf and high impedance). Prior to using anything but your voltmeter, check the power supply (perhaps the most important circuit in the transceiver). Before you undertake any other procedure, measure and verify that these voltages (regulated and unregulated) are where they are supposed to be.

Controlled Quality Factor (Q) Transistors

When, as a technician, you encounter statements such as "family of *controlled Q transistors,*" it can become confusing because the letter "Q" is used in the electronics field for many things; for example, quantity of electric charge expressed in coulombs; the identification letter for transistors in a schematic diagram; the ratio of reactance of a circuit to the resistance of a circuit; and a measure of sharpness or the frequency selectivity of an electrical system such as a transistor (the one we are discussing when we say *controlled Q transistor*).

Actually, what is desired in controlled Q transistor systems is that the input and output network presents a certain impedance (usually 50 ohms), and that the transistor will operate at the desired output power and efficiency over a given bandwidth. The bandwidth of Motorola's controlled Q transistor is about 450 to 512 MHz. As you probably know, a variety of police, taxi, trucking, and public utility maintenance communications systems operate in this part of the UHF spectrum.

Internal matching in the transistor is done by a transformer network (fabricated inside the transistor package) that is necessary to raise the normally low base impedance of the transistor to the desired 50 ohms, over the entire bandwidth of operation. As an example, in some designs the input/output filters are low-pass networks composed of a shunt capacitor, a series transmission line, and a shunt capacitor. The series transmission line is used in place of an inductor because it may be printed and, of course, reduces cost.

It is very important that you remember that any internal circuits must be coupled to a transistor. For proper coupling of two rf circuits, filter networks frequently are used to match one impedance value to another one and, as we have explained, these are often called *matching networks*.

The most important consideration in any coupling network you place between a transistor and a load is the amount of power delivered to the load over the desired bandwidth (Q of the circuit). The circuit you want to couple to the transistor has inductive and capacitive reactance, plus resistance. The reactive effects associated with even small inductance and capacitance place drastic limitations on UHF transistor operation if you do not "cancel them out" by placing the proper coupling network (inserting the right amount of the opposite kind of reactance) between the source (transistor) and a reactive load. It is beyond the scope of this book to cover coupling networks; however, most handbooks of practical electronics reference data and manufacturers' reference data will contain this information.

Another term we often encounter in the shop is *insertion loss*. The insertion loss is the ratio of the power delivered to the load with the coupling network in the circuit, to the power delivered to the load without the coupling network.

A Brief Review of Modern Resonant Circuits

Solid state devices such as tuning diodes (voltage-variable-capacitance diodes) are used as the variable element of a circuit to change (or tune) its *resonant* frequency. To review what resonance means, let's examine resonant circuits.

In general, all rf circuits (whether within an IC or designed using discrete components) are based on the use of resonant circuits consisting of some form of capacitance and inductance connected in series or parallel, as shown in Figure 4-4.

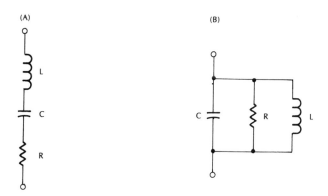

Figure 4-4: A series circuit is shown in (A). The circuit shown in (B) is parallel. Resonant frequency = $1/(2\pi \sqrt{LC})$, where L is in henrys, C is in farads, and frequency is in Hz).

At the resonant frequency (see Figure 4-4), the inductive and capacitive reactance (X_L and X_C) are equal, and the parallel circuit presents a high input impedance, or a low input impedance in the case of the series circuit.

However, in either circuit any combination of capacitance and inductance will have some resonant frequency, as the formula shows.

To permit tuning of the resonant circuit over a given frequency range, either the capacitance or the inductance can be variable. Therefore, when we replace the capacitance with a tuning diode, capacitance becomes the variable element in the resonant circuit. In any case, the two basic design considerations for rf resonant circuits are resonant frequency and the Q factor that was discussed in previous pages.

Usually, resonant circuit Q is measured at points on either side of the resonant frequency where the signal amplitude is down either 6 dB (0.707 × peak voltage or current), or down 3 dB (half-power), in respect to the resonant value of voltage or power. Figure 4-5 shows current in series-resonant circuits having different Qs (see Figure 4-5 A). Relative impedance of parallel-resonant circuits with different Qs is shown in Figure 4-5 B.

The formula for the Q of a circuit that has the various bandwidths shown in Figure 4-4 is

$$Q = F_0/(F_1 - F_2)$$

where F_0 is the resonant frequency and F_1 is the lowest frequency and F_2 is the

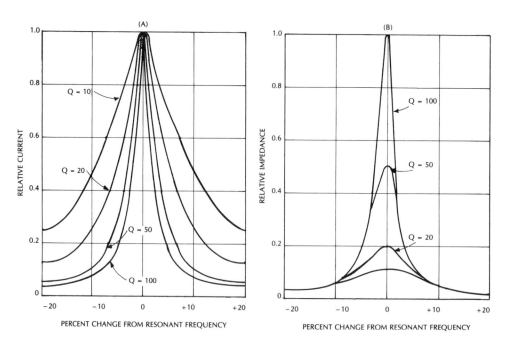

Figure 4-5: Series-resonant circuits having different Q factors but same peak current. *Note:* the lower value circuit impedance characteristic at different value Qs.

high frequency at the 3 dB bandwidth points. Or, another formula you can use is

$$\text{bandwidth} = (3 \text{ dB}) = F_0/Q.$$

If you work out a few example problems, you will find the Q must be increased for an increase in resonant frequency if you desire to retain the same bandwidth. Another point to remember is that you must decrease Q if you wish to increase bandwidth.

In summary, the input and output tuning circuit of an rf solid state amplifier must perform two functions:

1. Tune the amplifier to the desired frequency. Peak up at the resonant frequency while, at the same time, permitting the half-power bandwidth frequencies to pass at 0.707 × peak amplitude.
2. Match the input and output impedance of the transistor (for example, the control Q transistors we discussed previously, or any other rf transistor, for that matter) to the impedance of the rf signal source and load. If you do not do this, there will be a considerable loss of signal, due to the mismatch and resulting standing wave ratio and, of course, your tests and measurements will probably be way out of specs.

Servicing Practical Solid State RF Circuits and Devices

Regardless of the application—amateur radio transmitters, AM, FM transmitters, or single sideband CB transceivers—rf semiconductors are common elements in all the transmitters that most of us come into contact with today. These solid state devices are used in oscillators, rf amplifiers, frequency multipliers, and frequency down-converters. For example, one application of a down-converter is a home satellite TV system. This system converts satellite frequencies to the lower commercial broadcast frequencies needed by home TV receivers.

When you select a semiconductor device (such as an rf transistor), your choice will depend primarily upon the frequency band in which you want to operate, and the power output needed. For instance, a simple transistor oscillator circuit that meets your frequency stability needs could be used as a transmitter at some low rf frequency, but the power output would be quite low.

As a general rule, the output of a very stable oscillator (usually crystal-controlled) is fed into one or more amplifiers to bring the power up to the desired level. The last amplifier is an rf power amplifier, in every case. Figure 4-6 shows a working schematic of a single transistor rf power amplifier.

In this type of rf amplifier, the radio frequency chokes (RFC) should present an inductance reactance impedance of about 1 k ohm to 3 k ohms at the operating frequency. The bypass capacitors' value should be about 0.001

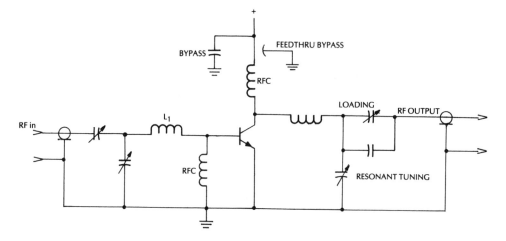

Figure 4-6: Typical discrete component rf power amplifier circuit. The selection of the transistor and components would depend on operating frequency and power input/output requirements of the transmitter design.

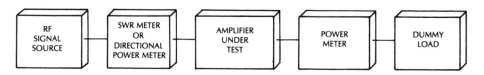

Figure 4-7: Test setup for adjusting an rf amplifier.

to 0.1 µf. Also, note that one of the capacitors in the output tuning circuit is labeled "Loading," and the other is "Resonant tuning." When tuning a circuit such as this, one capacitor (tuning) is tuned to the resonant frequency and the other (loading), is adjusted for proper impedance match to the load placed on the rf output line. To test an rf amplifier such as the one shown in Figure 4-6, you can use the test setup shown in Figure 4-7 for adjusting the amplifier.

The rf amplifier should be connected as in Figure 4-7. In general, you first adjust the input until you read minimum reflected power (SWR as low as possible), and then adjust the amplifier circuits until you read the highest output power possible on the power meter. This should be done with only a small amount of power being supplied by the rf signal source, and it will probably require tuning the amplifier output circuit capacitors several times. Also, in many cases a small cooling fan is needed to cool the power transistor heat sink during the test—especially if you are going to test for maximum power out. There are many rf power transistors available that you may encounter while servicing. Table 4-2 lists a few with different frequency and power ratings.

Device Type	P_{in} Input Power Watts	P_{out} Output Power Watts	G G_{PE} Power Gain dB Min	V_{CC} Supply Voltage Volts	Package
MRF466	0.1	3.0 PEP	15	12.5	TO-220
MRF406	1,25	20 PEP	20	12.5	211-07
MRF421	10	100 PEP	10	12.5	211-08
		1.5–30 MHz HF/SSB TRANSISTORS			
MRF476	0.10	0.5	10	12.5	TO-39
MRF 450A	4.0	50	11	12.5	145A-09
MRF458	5	80	12	12.5	211-11
		14–30 MHz, CB/AMATEUR TRANSISTORS			

Table 4-2: Low-voltage rf power transistors.

Radio Frequency Measurements

In the beginning of this section, it was shown that when dealing with electromagnetic waves it is often preferable to use wavelength rather than frequency; for instance, when dealing with optoelectronic devices. However, when working with rf circuits that are to be used with any of the rf power transistors listed in Table 4-2, it is usually more convenient to measure frequency rather than wavelength.

By and large, measurements in the rf spectrum require special instruments, and the frequency counter is probably the best all-around instrument offered today. These counters often provide frequency measurements in both UHF and VHF bands; in addition they may be hand-held types for in-the-field work, or they may be more sophisticated bench types for in-shop use. In most cases, these counters can be directly connected to the circuit under test. Or, for frequency counts without a direct connection, you can use an antenna to take a count of the frequency being radiated from the device under test.

Special RF Measurement Precautions

If you have not tried it, you will soon find out that radio frequency tests are quite a bit more demanding than audio frequency tests. Sales catalogs tell you all about how accurate their frequency counters are, but they don't tell you about problems you can encounter when using the instrument at the workbench; for example, how easy it is to get into trouble due to stray coupling between test leads or between circuit components; or how easy it is to lock on to a harmonic rather than the fundamental frequency. You will find that even the smallest value of stray reactance (for instance, a resistor wire lead can easily become a small inductor at UHF frequencies), is one of the factors that will make high frequency measurements more exacting. The following precautions will go a long way toward making your day at the workbench more pleasant when rf-testing:

1. Keep all wire leads as short, straight, and as large a diameter wire as possible.

2. Use a common ground point for all test instruments, circuits, etc. That is, avoid using a separate ground point for each rf system, whether it is a frequency counter, transmitter, receiver, or any other piece of the test setup.

3. For proper operation, use bypass capacitors and rf chokes (RFCs) in your circuit. The bypass capacitors and RFCs are usually called for in any rf amplifier circuit. See Figure 4-6.

4. Place your hand close to an unshielded oscillator circuit and you will see it change frequency of operation (assuming that you have a frequency-measuring instrument connected to an operating rf circuit). This is called *body capacitance,* and it can detune or reduce signal levels in an unshielded rf circuit. The problem may be caused by improper grounding and/or inadequate shielding.

5. If you are experiencing *drift* (a slow change in frequency), it is probably caused by trying to test equipment before it has had time to warm up properly. The equipment under test, and all test instruments, must be at the same temperature (usually room temperature) and *stabilized* at that temperature.

6. Another common problem is *electrical noise pickup.* This perplexing problem can be held to a minimum if you always use *short leads, proper power line filters,* and *rf shielding.* In some cases where radio interference is a problem, your best bet is to work in a shielded booth, called a *screen room.* This is a small room completely inclosed (sides, top, and bottom) with a small mesh metal screen. However, today's counters are generally housed in an interference-free cabinet so that such extremes are not necessary.

7. Don't overdrive the circuit under test. This problem can be prevented by using loose coupling between the circuit under test and test instrument. For best results, always use the loosest practical coupling. For example, if you are using a frequency counter with antenna attached, vary the distance between the counter antenna and the circuit you are testing until you have the least signal pickup you can have and still be able to make a frequency count. This will help assure that you are measuring the fundamental frequency, and improve the accuracy of your measurement.

8. You should calibrate rf test instruments at regular intervals. Many instruments have self-contained frequency standards that can be used for calibration.

A Collection of Formulas Frequently Needed in RF Work

Conductance, Susceptance, and Admittance Formulas

$$\text{conductance } G = 1/(R^2 + X^2)$$
$$\text{susceptance } \beta = 1/X_L \text{ (when } R = 0)$$
$$\text{admittance } \gamma = 1/\sqrt{R^2 + X^2}$$

Capacitance and Inductance: Reactance Is Known

$$C = 1/2\pi F X_C \text{ and } L = X_L/2\pi F$$

Power Factor Formulas (pf)

$$pf = \cos \theta, \text{ where } \theta \text{ is the angle of lead or lag}$$
$$pf = \text{watts/(current)(voltage)}$$

Inductance and Capacitance Needed for Resonant Frequency when Frequency and One Value Are Known

$$L = 1/(4\pi^2)(Fr^2)C$$
$$C = 1/(4\pi^2)(Fr^2)L$$

Impedance Formulas

$$Z \angle\phi = \sqrt{R^2 + X^2} \tan^{-1} X/R \text{ (for series circuit)}$$
$$Z \angle\phi = RX \sqrt{R^2 + \quad X^2} \tan^{-1} R/X \text{ (for R and X in parallel)}$$

Decibel Formulas

$$dB = 10 \log P_2/P_1 = 20 \log E_2/E_1$$
$$= 20 \log I_2/I_1 \text{ (when input/output impedance are equal)}$$
$$dB = 10 \log P_2/P_1 = 20 \log E_2 \sqrt{Z_1}\big/E_1 \sqrt{Z_2}$$
$$= 20 \log I_2 \sqrt{Z_2}\big/I_1 \sqrt{Z_1} \text{ (when input/output impedances are not equal)}$$

Frequency Modulation Index

$$\text{modulation index} = \text{frequency deviation/modulation frequency.}$$

SECTION 4.2: TESTS AND MEASUREMENTS NEEDED TO SERVICE RADIO RECEIVERS

This section includes solid state (both discrete and integrated circuits) and tube-type radio receiver tests and measurements, and concentrates on the easiest and most practical way to make them with a minimum of equipment. However professional work hasn't been forgotten. In many sections, you'll find several methods that are given in progressively greater detail, with more sophisticated test gear, that will insure optimum results for even the most demanding situation.

Whether you are an experienced technician interested in testing radio receivers more efficiently, or you're a beginner, this section will provide the information you need for specific types of tests and the specific test equipment you'll need to make each particular test using timesaving techniques. In every case, you may go directly to the test or measurement that describes the test procedure for a particular circuit or complete unit (use the Table of Contents or Index to locate the desired test). The tests and measurements are complete by themselves, with only an occasional reference to other tests.

Checking the Local Oscillator—Transistor Receiver

Test Equipment:
VTVM or equivalent type voltmeter

Test Setup:
See Figure 4-8.

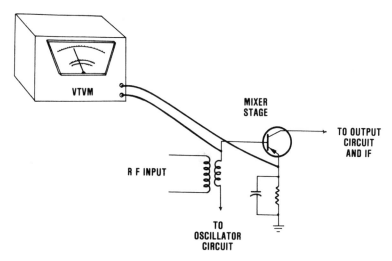

Figure 4-8: Test equipment setup for checking a transistor receiver local oscillator with a voltmeter.

Comments:
An applied test signal is not required when checking a receiver's local oscillator because it is a self-regenerating circuit.

Procedure:
Step 1. Attach a high input impedance voltmeter between the emitter and base (See Figure 4-8).

Step 2. It is not important what value voltage you read. All you want is a reading. (Of course, the reading will be quite low—a few tenths of a volt). If the voltage remains fixed, the oscillator is not operating. If it changes a few tenths of a volt (increase and decrease as you tune), the oscillator is operating.

Checking a Local Oscillator—Tube Receiver

Test Equipment:
VTVM or equivalent type voltmeter

Test Setup:
See Figure 4-9.

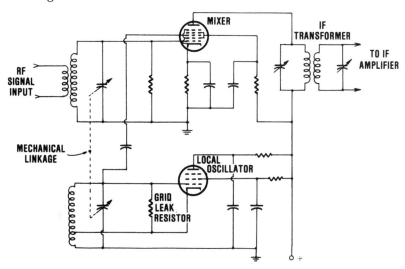

Figure 4-9: Pentagrid mixer with separate local oscillator. Many times you'll find the pentagrid converter acting as both local oscillator and mixer. However, it will have a slightly different wiring diagram.

Comments:

Many times you'll find that an AM radio uses a pentagrid tube and local oscillator to make up the mixer. Three formulas that are very handy when servicing radio receivers are: LO = RF + IF; RF = LO − IF; and IF = LO − RF: where LO is the local oscillator frequency, RF is the radio frequency being fed to grid number one of the pentagrid converter, and IF is the intermediate frequency of the receiver. In almost every case, you'll find the LO and RF frequencies are beat together and the difference is 455 kHz (the IF).

Procedure:

Step 1. Connect the voltmeter hot lead to chassis ground (remember, we're measuring tube bias which, normally, is negative in respect to ground). Set the range switch to 10 Vdc.

Step 2. Connect the voltmeter ground lead to the oscillator grid number one (See Figure 4-9). *Note:* If your voltmeter has a negative-position lead switch, merely set the switch to the negative position and make the connections.

Step 3. When the oscillator is operating, there is a negative voltage developed across the grid resistor because of the grid leak bias design. For the same reason, when the oscillator is not operating there is no voltage developed across the grid resistor. Therefore, if you measure zero volts on the voltmeter, the oscillator is "dead." On the other hand, if you measure about 5 volts dc, plus or minus 3 volts, the oscillator is working.

IF Amplifier Oscillation Tests

Voltmeter Method

Test Equipment:
   dc voltmeter

Test Setup:
   Connect a dc voltmeter across the load resistor of the receiver's second detector.

Comments:
   IF amplifier oscillation is caused by regeneration (regenerative feedback) which, in turn, may be caused by an open neutralizing capacitor or improper type of transistor being used as a replacement. The symptom most frequently noticed is a distorted receiver audio output signal with poor intelligibility, particularly when tuned to a weak station. In severe cases, the receiver may not have any output signal or you may hear a high squeal.

Procedure:
   Step 1. Connect your voltmeter, set to read dc, across the second detector load resistor.
   Step 2. If the receiver is nonoperative and you read a high dc voltage across the resistor, it's a sure sign there is an oscillating IF amplifier. When the receiver is operating but has a distorted audio output, refer to the service manual or make a comparative check with a properly operating receiver of the same type to determine the correct voltage. If the IF amplifier is oscillating, you'll find a higher dc voltage present at the detector output.
   Step 3. To make a positive check, measure the bandwidth of the suspected IF amplifier. The sign of regeneration with this test is a comparatively narrow response curve. If the receiver oscillates with the AVC clamp voltage removed but stops with the clamp voltage connected, you probably have an open bypass capacitor on the AVC line.

Signal Generator Method—Transistor Radio

Test Equipment:
   RF signal generator, jumper lead and trimmer capacitor (10 - 350 pF)

Test Setup:

Connect the signal generator to the receiver antenna by using a few turns of wire. Set the generator to a frequency of about 1 MHz.

Procedure:

Step 1. Turn on the signal generator and receiver after you have the setup completed. Adjust the receiver volume control for a convenient level. Assuming that an IF amplifier is oscillating, you may hear a squeal from the receiver at this point. However, there are other symptoms (see the preceding test).

Step 2. Take the jumper lead and short out the IF transistor you suspect, between emitter and base. When you do this, it cuts the transistor off by "killing" the forward bias.

Step 3. Connect the variable capacitor from the bottom of the IF amplifier output transformer to the base. See Figure 4-10.

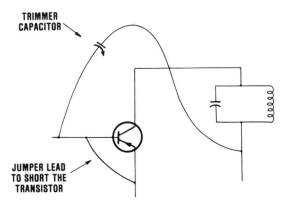

Figure 4-10: How to connect a trimmer capacitor during an IF amplifier oscillation test.

Step 4. Adjust the trimmer capacitor until you have minimum sound. If you can reduce sound to very little, or no sound, it's an indication that the IF amplifier was oscillating. Possibly the neutralizing capacitor is open. In this case, place a fixed capacitor of the same value as the trimmer capacitor in the circuit. Or, possibly, you'll want to leave the trimmer capacitor in the circuit as is (assuming the original capacitor is open).

AM Radio Antenna Test—Ferrite Core

Test Equipment:

None

Test Setup:

None

Comments:

This test is to find the best position for the antenna windings on the ferrite core of an AM radio receiver. Figure 4-11 is an illustration of an AM radio ferrite core antenna.

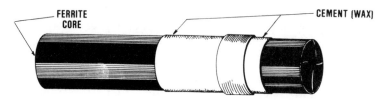

Figure 4-11: Antenna windings on an AM radio receiver ferrite core antenna.

Procedure:

Step 1. Free the antenna windings that you'll find wrapped around the ferrite core, with cement solvent, or loosen the wax. You can do this with your finger.

Step 2. Tune the receiver to the low end of the dial (any weak station near 600 kHz) and slide the antenna windings back and forth until you have the best possible signal.

Step 3. Tune the receiver to the high end of the dial (any weak station near 15 kHz) and adjust the antenna trimmer capacitor for maximum volume.

Step 4. Repeat Steps 2 and 3 until the receiver is operating at peak volume at both ends of the receiver tuning dial. Then re-cement the antenna windings to the ferrite core. It may appear that there has been no change in the antenna coil position. However, even the smallest movement can make a considerable improvement in reception, especially in weak signal areas.

Antenna Test Using a Dip Meter

Test Equipment:

Dip meter capable of tuning the frequency range of the receiver.

Test Setup:

Magnetically couple the dip meter to the receiver antenna.

Procedure:

Step 1. Tune the receiver to the low end of its dial.

Step 2. Tune the dip meter to the same frequency and look for the meter dip. If you get a good strong dip, it means the antenna is operating properly at this frequency.

Step 3. Tune the dip meter to several other frequencies in the band the receiver is set to. Adjust the dip meter to the same frequency each time and look for the greatest dip (it's possible to tune up to a harmonic, so be sure to *look for the strongest dip*). If you get good dips at each test frequency, the antenna is good. If not, check the antenna connections, lead-in, and other components in the antenna system.

Measuring Receiver Sensitivity

Power Method

Test Equipment:
    Signal generator and voltmeter

Test Setup:
    See Figure 4-12.

Comments:
    Basically, what you're doing when you make a receiver sensitivity measurement based on power output is making an overall receiver gain test without using internal noise as a reference. This measurement simply determines what level modulated signal is required at the receiver input to produce a certain power output (usually one-half watt, in better communications receivers).

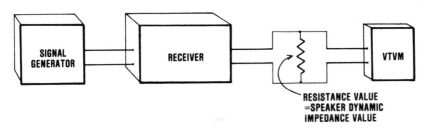

Figure 4-12: Test setup for radio receiver sensitivity measurement.

Procedure:

Step 1. Connect the equipment as shown in Figure 4-12 and set the receiver rf and AF gain to maximum.

Step 2. Adjust the signal generator to produce 1.4, 2, etc. volts across the load resistor. For all practical purposes, this is equal to 0.5 watts, assuming a load resistor of 4 or 8 ohms (using the formula $P = E^2/R$).

Step 3. Measure the RF voltage into the receiver. This value is the sensitivity of the receiver. Notice that no signal-to-noise ratio (S/N) is included. However, this check is very good to keep a running account of a receiver's performance or to compare one receiver with another.

Precision Signal Generator Method

Test Equipment:

High input impedance voltmeter, RF signal generator and dummy antenna, if needed

Test Setup:

See Figure 4-12.

Comments:

Sensitivity is expressed as the signal level required to produce a specified output signal having a specified signal-to-noise ratio. Typically, it is the signal input needed to produce an output 10 dB above the receiver's internal noise.

To make a meaningful sensitivity measurement on a high-gain communications receiver, the signal generator must have a good metering device. This is particularly important when working with high frequencies. Also, the voltmeter used during the measurement must be sensitive enough to read the receiver's internal noise when it is connected across the load resistor.

Procedure:

Step 1. Tune the receiver to a position on the dial where no station is detected. You should hear only noise with the gain control set at maximum.

Step 2. Remove power and disconnect the speaker from the receiver output and connect a resistor of the same impedance value as the speaker, in place of the speaker.

Step 3. Apply power and connect a voltmeter across the load resistor. If the receiver uses a 4-ohm speaker, you may have trouble reading a voltage across the resistor. In this case, and assuming there is an output transformer, move the voltmeter connections to the output transformer input winding, i.e., the transformer primary winding.

Step 4. A multiple band receiver should be set to its lowest frequency bandwidth and the RF sensitivity set to maximum.

Step 5. Adjust the audio volume control and voltmeter range switch until the VTVM reads 0 dB on the meter scale. This is your reference point with no signal input from the signal generator.

Step 6. Using a shielded cable, connect the signal generator to the receiver input. If the receiver usually is connected to a 50-ohm antenna, you can attach the signal generator output test lead directly to the receiver antenna input, because most signal generators have a 50-ohm output impedance. However, if the receiver has some other input impedance, you'll have to use a suitable dummy antenna. Several different types are shown in Figure 4-13. To select the correct type to use, you should consult the receiver service manual.

Dummy antennas are easy to make, but keep your leads as short as possible and mount the network in a metal box. An rf dummy load must be

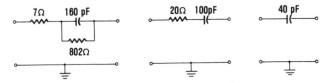

DUMMY ANTENNA FOR AUTOMOBILE RECEIVER TESTS
(CONSULT THE SERVICE MANUAL)

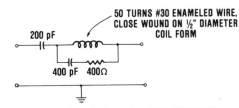

DUMMY ANTENNA FOR TESTING RECEIVER SETUP THAT WORKS INTO
A RANDOM LENGTH ANTENNA

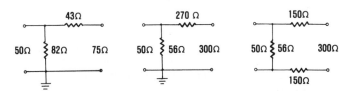

DUMMY ATENNA FOR MATCHING IMPEDANCES SHOWN
(CONSULT THE RECEIVER SERVICE MANUAL)

Figure 4-13: Receiver testing antenna networks.

well shielded to prevent stray rf field pickup that can cause erroneous readings.

Step 7. Typically, signal generators are manufactured to produce 30 percent modulation. However, if you are using external modulation or your signal generator has a variable internal modulation control, set it for 30 percent modulation at the frequency of the receiver.

Step 8. Next, adjust the signal output level of the signal generator to 1/10 of a volt (0.1V) with the signal generator step attenuator set at minimum attenuation. If your signal generator doesn't have a metered output, use a high input impedance voltmeter to make the setting.

After you've made the 0.1V setting, do not change the signal generator's variable attenuation control knob. At this point, all changes of output level must be done by changing the signal generator step attenuator. You should now have an output level of 0.1V with the signal generator step attenuator set at zero dB, and the variable attenuator control set to maximum output.

Step 9. The step attenuator of a precision rf generator is frequently calibrated in dB. In this case, you have to convert the dB settings to microvolts. With steps divided by 10 and a reference of 0 dB = 0.1V, we have the result shown in Table 4-3.

0 dB	=	0.1 VOLT
20 dB	=	0.01 VOLT
40 dB	=	0.001 VOLT
60 dB	=	0.0001 VOLT
80 dB	=	0.00001 VOLT
100 dB	=	0.000001 VOLT
	=	1 MICROVOLT = 1μ VOLT

Table 4-3: Conversion of dB's to volts output when signal generator output attenuator is calibrated in dB's.

Step 10. With the step attenuator set so that the output voltage is 0.001 or 0.00001 volts (60 or 80 dB), adjust the signal generator frequency for a maximum reading on your voltmeter. If your voltmeter reading is off scale, increase the attenuation.

Step 11. The gain is read directly off the step attenuator, if the measurements fall on even numbers. The total attenuation required to reduce the output to your reference is the gain in dB. If the attenuator reading doesn't fall on even numbers, adjust the variable attenuator knob on the signal generator for a reading on your voltmeter that is as close as possible to 10 dB below the zero noise level. If the receiver output meter reads 13 dB, for instance, reduce the signal generator variable control to the 10 dB point. Notice this is 3 dB down. Therefore, using the formula, dB = 20 log voltage$_2$/voltage$_1$, we get, 3 = 20 log 0.707 or, approximately, 0.7. Multiplying this value by our reference (1μV), we get 0.7 μV. This means the sensitivity of the receiver is 0.7 μV for 10 dB S/N.

Should the nearest setting be below 10 dB, you will have to turn the variable control on the generator up to where the receiver output meter reads 10 dB above zero. The voltage you read on the signal generator output, multiplied by your reference in μV, will give you the sensitivity of the receiver for 10 dB S/N.

Audio Stage Gain Using a Voltmeter and AF Signal Generator

Test Equipment:

AF signal generator, high input impedance voltmeter with demodulator probe, and a load resistor to replace the speaker

Test Setup:

Connect the AF generator set at 400 Hz (if you use 1,000 Hz, the following example will change values), across the second detector (AM demodulator) load resistor. Replace the speaker with a resistor (4, 8, or 16 ohms), capable of dissipating the receiver output load without changing value. Connect an ac voltmeter across the resistor you use to replace the speaker.

Procedure:

Step 1. Connect the equipment as was described under the heading **Test Setup.**

Step 2. Adjust the audio signal generator output to between one-half and one volt.

Step 3. Measure the voltage across the receiver terminating resistor.

Step 4. Let's say that the receiver is rated to deliver 2 watts to a load impedance of 8-ohms. A reading of 4 volts should be obtained with the one-half to one-volt input to the audio stages. To calculate the correct output voltage reading for various load resistors and powers, simply use the formula: voltage out = $\sqrt{(\text{resistance})\,(\text{rated power})}$.

IF Stage Gain Measurement

Voltmeter and RF Signal Generator Method

Test Equipment:

RF signal generator, high input impedance voltmeter, demodulator probe and dummy antenna, if needed

Test Setup:

Set the voltmeter to measure a dc voltage and connect the demodulator probe across the AF demodulator (AM detector) load resistor. Set the range to a low scale, 0 - 1.5 volts, for example. Connect the signal generator to the input of the last IF stage.

Comments:

The gain of an IF stage depends upon its bandwidth. For instance, if the bandwidth of the stage is doubled, it will lose about one-half of its gain. Therefore, it would be impossible to achieve the gain given in the manufacturer's specs unless the bandwidth were set exactly as the service manual recommends. Reference to the service notes is the best method to find the correct gain for the receiver under test. Another good way is to make a comparative check in a normally operating receiver of the same type.

The overall gain of an IF amplifier strip is equal to the product of the individual identical stage gain. If three identical stages are used on the IF strip, the overall gain is the cube of the gain of a single stage. For example, ideally, a triode transistor connected in a common emitter configuration will produce a gain of 270. Therefore, three identical stages will produce a voltage gain of approximately 270^3. With all this gain, the output of the last IF amplifier still will be only somewhere in the vicinity of a few tenths of a volt and, as you can see, it's all but impossible to measure the preceding stages with anything but a very good instrument.

Procedure:

Step 1. Connect the signal generator to the input of the last IF stage. If a dc blocking capacitor is needed, place a 0.01 μF capacitor in series with the signal generator's ungrounded test lead.

Step 2. Set the signal generator to the intermediate frequency of the receiver and adjust the variable output level control until you read 0.1 volts on the voltmeter.

Step 3. Move the signal generator test lead to the output of the last IF stage.

Step 4. Adjust the signal generator attenuator until you have a reading of 0.1 volt on the voltmeter. The difference in the readings is the voltage gain of the stage.

Measuring IF Stage Gain in dBs Method

Test Equipment:

RF signal generator with its attenuator calibrated in dBs, high input impedance voltmeter and demodulator probe

Test Setup:

Connect the voltmeter demodulator probe across the receiver demodulator (AM detector) load resistor with the voltmeter set to measure dc volts. Set the range switch to a low range; 0 to 1 volt, for example. Connect the signal generator to the input of the last IF stage. If a dc blocking capacitor is needed, use a 0.01 μF capacitor in series with the generator's ungrounded test lead.

Procedure:

Use the same method employed in the preceding test, except that when the signal generator attenuator is calibrated in dBs the gain of a stage is read directly off the attenuator range setting, if the measurements fall on even numbers. In cases that fall in between the even dB readings, use the generator's variable attenuation control. The total attenuation required to reduce the output to your reference is the gain of the stage in dB. However, since gain frequently is a voltage ratio, it may be necessary to convert. There is a voltage gain of 10 for each 20 dB (See Table 4-3). Incidentally, this measurement is basically the same as described in the Precision Signal Generator Method.

IF Stage Gain Measurement—Oscilloscope Method

Test Equipment:
RF signal generator, oscilloscope, high frequency probe

Test Setup:
Connect the signal generator through a 0.01 μF dc blocking capacitor or a small coil (a few loops of insulated wire will do), to the antenna of the receiver. Connect the scope probe to the input and output of the stage under test, as explained under **Procedure.**

Comments:
This test calls for a high frequency probe and it's important that you do not use a probe with excessively high capacitance. Many "home brew" and low-cost probes are simply small trimmer capacitors. In this case, adjust the probe capacitor to as small a value as possible to produce on the scope (as near as possible) a perfect reproduction of the signal generator output. Too much capacitance will detune an IF transformer and cause erroneous readings. Also, don't overload the amplifier.

It is important to stabilize the gain of a receiver before making the test. In most cases, you'll find the gain of an IF stage depends on the AVC bias voltage developed at the second detector and fed back to the base of the IF amplifier transistor circuit. This is a variable voltage and should be set at a standard AVC clamp voltage; for example, 1.5 volts. See the service manual for the value needed for the particular receiver you're testing.

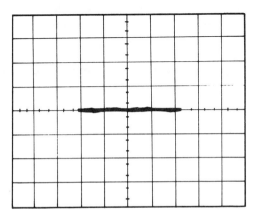

Figure 4-14: Oscilloscope display with high frequency probe connected to an IF amplifier input circuit.

Procedure:

Step 1. Let all equipment warm up for about one-half hour.

Step 2. Connect the rf signal generator test lead to the receiver antenna.

Step 3. Connect the scope high frequency probe to the input of the IF stage under test. You should see a waveform on the scope similar to the one shown in Figure 4-14.

Step 4. Move the high frequency probe to the amplifier output circuit. The rf signal generator signal should appear larger and appear on the scope as shown in Figure 4-15.

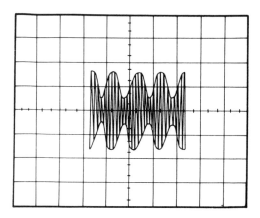

Figure 4-15: Oscilloscope display with high frequency probe connected to an IF amplifier output circuit.

Step 5. Adjust the scope attenuator until you have the second waveform (See Figure 4-15) set to the same amplitude as the first one (See Figure 4-14). If the attenuator is calibrated in 10s and you make two steps in attenuation, the gain of the amplifier is 20. Remember, gain is a measured voltage ratio and 20 dBs is a voltage ratio of 10 (See Table 4-3).

Measuring Local Oscillator Frequency Using a Frequency Generator and Scope

Test Equipment:
Oscilloscope and calibrated signal generator

Test Setup:
See **Procedure.**

Comments:
Sometimes, when a receiver's local oscillator is running off frequency, it will result in a "dead" receiver. Or, if the local oscillator is drifting, you will have to keep retuning the receiver. It's very easy to check the local oscillator

frequency and its stability with a signal generator and scope, as explained in the following procedure.

Procedure:

Step 1. Couple the scope probes to the local oscillator tuned circuit and adjust the scope until you see 1 to 4 cycles. Count the number of peaks in the waveform.

Step 2. Move the scope probe and connect it directly to the output terminals of the signal generator.

Step 3. Tune the signal generator until you see exactly the same number of cycles on the scope as you saw in Step 1. The reading you have on the signal generator dial is the frequency of the receiver's local oscillator.

Step 4. All you have to do to make a frequency drift test is monitor the scope pattern. The oscillator drift is determined by how much you have to adjust the signal generator to maintain the same number of cycles on the scope, and how often you have to make the adjustments. For example, if you have to correct five cycles per hour, that is the oscillator drift per hour.

Automatic Volume Control Tests

AVC Voltage Test—VOM Method

Test Equipment:
Voltmeter

Test Setup:
Connect the VOM between the AVC line and chassis ground (or power supply negative terminal).

Comments:
Automatic volume control (AVC) is a self-acting device that is supposed to maintain the output of a radio receiver or amplifier at a substantially constant level within relatively narrow limits, while the input voltage varies over a wide range. This same definition describes the operation of *automatic gain control* (agc). An AVC system regulates the gain of the rf and IF stages and its output has to be a dc voltage. In a small transistor radio, it will be either negative or positive, depending on the type of transistor and which lead it is applied to: negative, for the PNP transistor, and positive, for an NPN, if the AVC is applied to the emitter. When the AVC is connected to the base of an NPN transistor, it should be negative, and for a PNP, it's just the opposite, positive.

It's very easy to check whether the AVC is supposed to be positive or negative. Simply look at the second detector diode polarity. If the cathode is connected to the AVC line, it's positive. If the anode is connected to the AVC line, it's negative.

A common AVC trouble symptom is distortion on strong stations. There are several ways to check out the AVC system with nothing but a voltmeter. In general, the higher the ohms-per-voltmeter you use, the more accurate your measurements will be. Also, the most accurate reading will be obtained on the highest usable range of the voltmeter. However, when servicing transistor radios, in some instances you'll have to use the voltmeter's lowest dc voltage range. For example, an AVC problem is indicated where you measure an AVC voltage of about 0.5 volts, if you measure it across the rf amplifier emitter resistor. Tube-type receivers will have much higher AVC voltages—somewhere in the 0 to 10 volt region.

Procedure:

Step 1. Set the VOM dc switch to the highest usuable dc range.

Step 2. In communications receivers that have a manual AVC switch, check to see that the switch is set to its AVC position.

Step 3. You must refer to the manufacturer's service data to make an ideal test. However, the AVC output voltage is supposed to be a dc voltage (if it isn't, you'll have distortion in the rf and IF signals). Monitor the VOM and tune the receiver to various stations. As the signal increases in strength, the gain of the controlled stages is thereby reduced by the increasing value of bias. Therefore, you should see a change in the dc voltage readings as you slowly tune from station to station. Incidentally, if the AVC filter capacitor should open, the AVC bias voltage will follow the rise and fall of the audio envelope and this may be heard as distortion in the receiver output.

Step 4. Finally, if you measure different AVC dc voltages on adjoining ranges of your VOM, it's an indication the VOM is loading the circuit under test. This can result in considerably different voltage readings from those recommended by the manufacturer. If you encounter this problem, it's possible to observe two of the different AVC voltage readings and use the formula:

$$E = [E_1 E_2 (R_2 - R_1)]/(E_1 E_2 - E_2 R_1)$$

where E is the true AVC voltage, E_1 is the VOM reading on the first range setting, R_1 is the VOM input impedance for the first range setting, E_2 is the VOM reading on the second range setting, and R_2 is the VOM input impedance for the second range setting. Or, you can use the following procedure.

AVC Voltage Test—Half-Voltage Method

Test Equipment:

VOM and 5 k-ohm variable resistor

Test Setup:

See Figure 4-16.

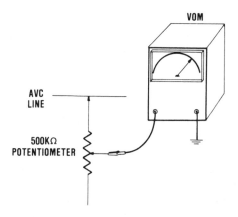

Figure 4-16: Measuring AVC voltage using the half-voltage method.

Procedure:

Step 1. Connect your VOM and a pot, as shown in Figure 4-16, and then set the VOM for about midscale, with the pot set at the zero resistance position.

Step 2. Adjust the pot resistance until the voltage is *exactly* one-half the reading you had in Step 1. Remove the pot and measure the resistance.

Step 3. Calculate the actual AVC voltage by using the formula:

$$E_{avc} = (E_1R_p)/R_{meter}$$

where E_{avc} is the actual AVC voltage, E_1 is the first voltage measurement, R_{meter} is the ohms-per-volt rating of the VOM times the full-scale voltage, and R_p is the resistance value you measured across the pot that was required to reduce the voltage to one-half its previous value. For example, let's say that you are using a VOM with an input resistance of 20,000 ohms-per-volt, and it's set to the 10-volt dc range. Next, assume that you measure 4 volts on the initial voltage reading, and a potentiometer resistance of 300,000 ohms when the VOM voltage reading has been reduced to 2 volts. The true AVC voltage is:

$$E_{avc} = (E_1R_p)/R_{meter} = (4)(300,000)/200,000$$
$$= 1,200,000/200,000 = 6 \text{ V}$$

AM Receiver Tracking Test

Test Equipment:

RF signal generator, ac voltmeter and dummy antenna (use the type specified by the manufacturer's service notes). See Figure 4-13 for typical dummy antennas used in radio receiver servicing.

Test Setup:

Connect an ac voltmeter across the speaker terminals. Connect an AM signal generator to the receiver antenna through a dummy antenna, as required by the manufacturer. Set the receiver volume control to maximum.

Comments:

Tracking is the maintenance of proper frequency relationships in circuits designed to be simultaneously varied by ganged operation. For example, the mixer and local oscillator circuits of a superheterodyne radio receiver are said to *track* if they maintain a constant frequency difference (usually the intermediate frequency equals the local oscillator frequency minus the input signal from the RF circuit) throughout the receiver tuning range. While it is impossible to receive all stations with equal sensitivity (commonly, stations are weaker at the top end of the tuning dial, or the middle of it), the receiver will be operating at peak performance once you have the tuning circuits tracking as closely as possible.

Procedure:

Step 1. Use a suitable dummy antenna and connect the signal generator to the receiver antenna. Set it to the high end of the dial—about 1400 to 1615 kHz. Some technicians use a standard fluorescent light. The light is an all-frequency signal source that will provide static anywhere you tune the dial. Simply make your tracking adjustments for maximum static in each of the listed steps. Weak radio stations also will do quite well, as explained at the end of this procedure.

Step 2. Using a weak test signal (with a fluorescent light, increase and decrease the noise level by varying the distance between the light and receiver), adjust the oscillator, rf, and antenna trimmers for maximum output. It's best to record your readings if you expect to check the receiver periodically because you can refer to them at some future date to see if the system is deteriorating.

Usually, the oscillator trimmer is in parallel with the oscillator tank and the antenna trimmer is in parallel with the antenna tuned input circuit. In automobile radio receivers, the oscillator tuning slug generally is set to some certain depth (see the manufacturer's service notes). As an example, 1-3/8 inches frequently is used as the distance from the back of the coil to the tuning slug with the receiver set to the high end of the dial.

Step 3. Turn the signal generator dial to 600 kHz and tune the receiver for maximum output.

Step 4. Adjust the signal generator to a weak signal level output and tune the antenna and rf slugs for maximum receiver output. Perform the adjustments a few times, for best results.

For a quick tracking check, set the receiver to some weak station at about 600 kHz, with maximum volume. Adjust the *oscillator* trimmer for maximum signal level. Next, set the receiver to a weak station somewhere in the high end of the band. Adjust the *antenna* trimmer for maximum signal. Do this a few times and it should do the trick. If it doesn't, you probably need a more sophisticated tracking test, like the one described in Steps 1 through 4. See Figure 4-17 for typical locations of antenna, rf, and oscillator tuning slugs in a mobile receiver.

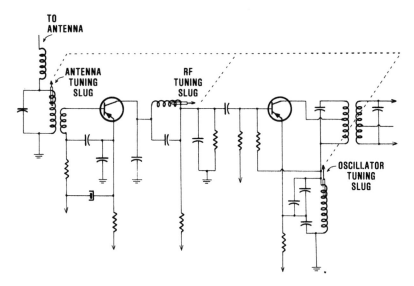

Figure 4-17: Typical location of antenna, rf, and local oscillator tuning slugs in a mobile receiver.

FM Demodulator Test

Test Equipment:
Sweep generator and oscilloscope

Test Setup:
Connect your scope across the volume control and inject the sweep generator signal into the limiter if the receiver is using a discriminator, or into the last IF amplifier if it has a ratio detector.

Procedure:

Step 1. Set your sweep generator to the receiver IF. Some receivers have an IF of 10.7 MHz and a bandpass of 200 kHz, but not all of them. In some cases, low-cost ones have a bandpass of only 50 kHz. Therefore, when in doubt, check the service notes.

Step 2. Disable the local oscillator by placing a jumper wire across the oscillator coil

Step 3. Adjust your scope sweep frequency to 60 Hz. Adjust the sweep generator to sweep 200 kHz.

Step 4. Adjust the demodulator by tuning the slug in the transformer secondary until the curve is equally divided on both sides of the vertical line on the face of the scope.

Step 5. Adjust the primary slug until you have an S-shape curve at about a 45° angle, something like that shown in Figure 4-18.

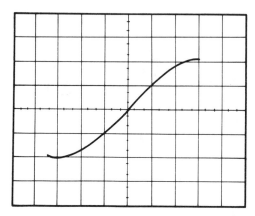

Figure 4-18: An example of the curve you should see on your scope when checking an FM demodulator.

Step 6. Set the horizontal sweep frequency of your scope to 120 Hz and you should see a new pattern on the scope. It should look like the one shown in Figure 4-19.

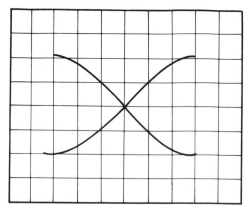

Figure 4-19: Scope pattern for last part of FM demodulator check.

Step 7. Adjust the secondary coil slug until the two curves form, as nearly perfect as possible, an X pattern like the one shown in Figure 4-19. A perfect X pattern is an indication the demodulator is operating properly.

IF Amplifier Test for FM Radio with Ratio Detector

Test Equipment:
Sweep generator and high input impedance VOM

Test Setup:
See Figure 4-20.

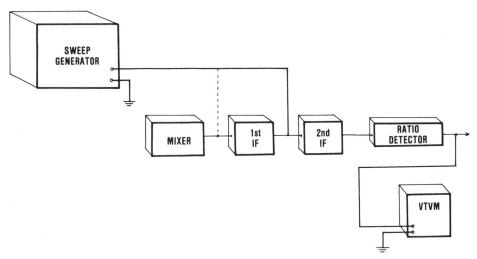

Figure 4-20: FM radio with ratio detector, IF amplifier test setup.

Comments:
Basically, this test is an IF alignment check.

Procedure:
Step 1. Attach the sweep generator to the first IF amplifier input circuit. Short the receiver local oscillator by placing a jumper wire across the oscillator coil.

Step 2. Set the sweep generator to the receiver IF frequency (normally 10.7 MHz).

Step 3. Set the sweep generator to the correct sweep frequency. This can be anywhere from 50 kHz up to more than 200 kHz, therefore, you'll have to refer to the manufacturer's service notes. As a rule of thumb, very inexpensive receivers will have about 50 kHz bandpass and very expensive ones will be from 150 to 200 kHz, with some all the way up to 250 kHz, but this much is rare.

Step 4. Connect the voltmeter across one side of the ratio detector. The output voltage may be positive or negative. Either one is okay because it doesn't make any difference which one you can get.

Step 5. Simply check each IF amplifier (in inexpensive FM radios, you'll find only two) and adjust for the highest reading you can get. This will produce the best response. If you can't get a decent signal strength reading during this test, it's an indication the receiver is in need of a more sophisticated alignment.

Stereo Amplifier Power Output Test

Test Equipment:
AC voltmeter or oscilloscope

Test Setup:
Connect the voltmeter or oscilloscope across the speaker coils.

Procedure:
Step 1. Set the gain controls of both channels to the same level.

Step 2. Measure the ac volts (rms) developed across each speaker voice coil with identical signal input. Remember, although the oscilloscope has the added advantage of letting you compare the two channels' waveforms, it reads a peak-to-peak voltage. Therefore, don't forget to convert the voltage measurement to rms voltage by dividing the peak-to-peak measurement by 2.8 before you do the next step.

Step 3. Compute the power out for each channel by using the formula, P $= E^2/R$, where R is the speaker impedance. Then make a comparison of the two. In most systems, they should be equal.

Step 4. To check the waveforms of each channel for distortion, use an identical input signal and monitor the output with your scope. A dual trace scope is ideal for this test.

Phase-Lock Loop FM Stereo Demodulator Adjustments

Quite a few new ICs are being manufactured by companies such as Motorola and you may encounter them when servicing modern radio receivers; for example, the phase-lock loop FM stereo demodulators MC1309 and MC1310P.

These devices require no inductors, have a low external part count, and only oscillator frequency adjustment is necessary. They also have a wide supply range (8 - 14 Vdc). In fact, to make a simple oscillator adjustment without equipment other than the stereo receiver itself is very easy. Tune the receiver to a stereo broadcast and adjust a variable resistor until a pilot light turns "on." To find the center of the lock-in range, rotate the potentiometer back and forth until the center of the lamp "on" range is found. This completes the adjustment. The result will be within a few dB of optimum.

SECTION 4.3: HOW TO SELECT AND USE HIGH FREQUENCY DIODES, TRANSISTORS, INTEGRATED CIRCUITS AND MODULES

In the past, electronic technicians needed to remember only a few practical semiconductor characteristics in order to service countless circuits for various applications. However, digital circuits and things such as personal computers have changed all of this. No longer can a few semiconductor devices be memorized and used in almost every situation.

Not only the semiconductor devices themselves, but also the circuits have changed over wide latitudes, for different applications. Thus, the need for this section is very evident to anyone who works with today's high-speed circuits, where cutting nanoseconds is the name of the game.

Servicing Electronic Tuning and Control Applications

As has been explained, the resonant frequency of a circuit can be shifted by changing either the inductance or the capacitance in the circuit. Also, a voltage-variable capacitive diode can be utilized in an oscillator circuit to shift the resonant frequency of the oscillator. To see how this can be done, let's examine a UHF ocsillator circuit in a piece of electronic gear we all know—a TV set. See Figure 4-21 for the schematic of a UHF television receiver tuner that is tuned by use of a voltage-variable diode.

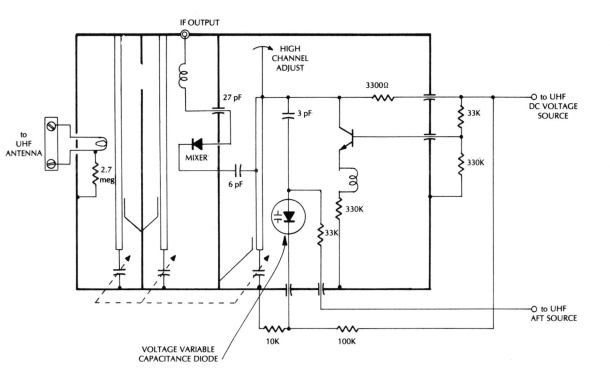

Figure 4-21: Schematic of a television receiver UHF tuner that uses a voltage-variable capacitance diode and UHF mixer diode.

Tuning Diode: The voltage-variable capacitance diode that may be called a *varactor diode* (especially if it is used in AM tuning), *silicon capacitor, voltage controlled capacitor,* etc., is a two-terminal solid state device whose capacitance varies with the applied voltage. Notice that this capacitance diode is in the UHF circuit and is utilized by the AFT (automatic fine tuning) circuit to adjust the UHF oscillator for optimum tuning. The AFT voltage applied to the capacitance diode determines the capacitance, and therefore the operating frequency of the UHF oscillator.

Mixer Diode: The *mixer diode* must also be a high frequency device because it must mix the incoming frequency (off the UHF television antenna) with the UHF oscillator frequency, to produce the intermediate frequency (IF) that is amplified in the TV receiver IF stages. As you can see from the schematic, this diode is also a two-terminal solid state device.

There are several different circuit arrangements that utilize voltage-variable capacitance diodes. However, tuning in these tuner circuits is accomplished by utilizing the characteristics of voltage-variable capacitor diodes. This causes them to behave as capacitors when they are reverse biased. The amount of reverse bias voltage applied to the diode will determine its capacitance.

It is now common to find these diodes used for tuning the antenna input, rf amplifier, mixer, and oscillator circuits. Most modern TVs and other receivers select a channel by use of push-buttons which apply the proper voltage for a particular channel to the voltage-variable capacitor diodes. The improvement in manufacturing of semiconductor devices is the primary reason behind the greatly improved performance of modern tuners, whether in TV, CB, or test instruments.

Hot-Carrier Diodes: Some of the names you will encounter when working with high frequency diodes are tuning, hot-carrier, and PIN diodes. We have discussed the tuning diode, so let's now take up the hot-carrier diode. The silicon hot-carrier diodes (Schottky barrier diode), MBD101 and MBD102, manufactured by Motorola, are designed primarily for UHF mixer applications but are also suitable for use in ultra-fast switching circuits.

These diodes (there are different packages. See Figure 4-22) have extremely fast turn-on and turn-off times, excellent diode forward and reverse characteristics, lower noise characteristics, and wider dynamic range. Typically, they have very low capacitance—less than 1.0 pF @ zero volts—and a low noise figure of −7.0 dB max. @ 1.0 GHz.

PIN Switching Diodes: A silicon switching diode may be in the same type of package illustrated in Figure 4-22. The anode and cathode pins will also be like those shown in that figure. However, a less expensive package looks like a small carbon resistor (body and two connecting leads); for example, the MPN3500 series, which is designed for VHF band-switching applications and general switching circuits.

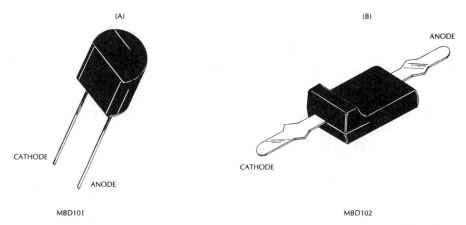

(A)　　　　　　　　　　　　　　　　(B)

ANODE

CATHODE

ANODE

CATHODE

MBD101　　　　　　　　　　　　　　MBD102

Figure 4-22: Silicon hot-carrier diode packages. (A) inexpensive plastic package. (B) low-impedance mini-L package for high volume requirements.

Notice that these devices will function properly at VHF band frequencies. The storage time for the PIN diode is long enough so that it completely fails to rectify at UHF (microwave) frequencies and above. When used in the microwave part of the rf spectrum, a PIN diode behaves like a variable resistor with its value controlled by a dc bias current. Therefore, you will find that it will work well as a variable microwave attenuator. In fact, there are several commercial microwave attenuators that are designed using PIN diodes. These devices are usually called *PIN-diode attenuators* and are a two-part network consisting of two or more PIN diodes controlled by a driven circuit.

SECTION 5

Practical Guide to In-Circuit Testing of High Frequency Transistors

SECTION 5.1: USING TRANSISTOR CHARACTERISTICS IN TESTS AND MEASUREMENTS

Most of the *needed* information for a particular transistor can be obtained from the data sheet that comes with the transistor when you purchase it. However, this is not true in every case. If you are working with extremely high frequency circuits, such as personal computer circuits, where high frequency switching characteristics are of particular importance, you may have to test a transistor using the actual working circuit you expect it to perform in.

Nevertheless, you will always have to refer to the data sheet to begin, because you will find that each manufacturer uses different ways to present the same data: maximum ratings and electrical characteristics. As you can imagine, it would be all but impossible to present all data sheet formats in this section (or in a book, for that matter); therefore we will use Motorola's presentation of transistor specifications.

Table 5-1 is the data sheet for an MPS-U02 that is a fairly typical amplifier transistor. For example, it has numerous substitute types; GE-63, TR-72, 2SC1384, PTC144, SK3199, ECG188, WEP3020, and RT-156, which are all general-purpose replacements.

This transistor is listed as an NPN *silicon annular amplifier transistor.* In case the word "annular" is new to you, it means a *mesa* transistor in which the semiconductor materials are arranged in concentric circles about the emitter. A mesa transistor is one in which the base and emitter semiconductor materials appear as plateaus above the collector region. A non-mesa semiconductor is called a *planar* device (for example, a planar diode).

Maximum Ratings

The first specifications listed in Table 5-1 are *maximum ratings.* The maximum collector emitter voltage is listed as V_{CEO}. Actually, this is a test

NPN SILICON ANNULAR AMPLIFIER TRANSISTOR

Designed for general purpose, high-voltage amplifier and driver applications.

High Power Dissipation — P_D = 100W @ T_C = 25°C

Complement to PNP MPS-U52

MAXIMUM RATINGS

Rating	Symbol	Value	Unit
Collector-Emitter Voltage	V_{CEO}	40	Vdc
Collector-Base Voltage	V_{CB}	60	Vdc
Emitter-Base Voltage	V_{EB}	5.0	Vdc
Collector-Current Continuous	I_C	800	mAdc
Total Power Dissipation @ T_A = 25°C	P_D	1.0	Watt
Derate Above 25°C		8.0	mW/°C
Operating and Storage Junction	T_j, T_{stg}	−55 to +150°	°C

THERMAL CHARACTERISTICS

Characteristic	Symbol	Value	Unit
Thermal Resistance, Junction to Case	$R_{\theta_{JC}}$	12.5	°C/W
Thermal Resistance, Junction to Ambient	$R_{\theta_{JA}}$	125	°C/W

ELECTRICAL CHARACTERISTICS (T = 25°C unless otherwise noted)

Characteristic	Symbol	Min	Max	Unit
OFF CHARACTERISTICS				
Collector-Emitter Breakdown V (I_C = 1.0mAdc, I_B = 0)	BV_{CEO}	40	—	Vdc
Collector-Base Breakdown V (I_C = 100μAdc, I_E = 0)	BV_{CBO}	60	—	Vdc
Collector Cutoff Current (V_{CB} = 40Vdc, I_E = 0)	I_{CBO}	—	100	nAdc
ON CHARACTERISTICS				
DC Current DC Current Gain	h_{FE}			—
(I_C = 10mAdc, V_{CE} = 10Vdc)		50	—	—
(I_C = 150mAdc, V_{CE} = 10Vdc)		50	300	—
(I_C = 500mAdc, V_{CE} = 10 Vdc)		30	—	—
Collector-Emitter Saturation (I_C = 150mAdc, I_B = 15mAdc)	$V_{CE(sat)}$	—	0.4	Vdc
Base-Emitter Saturation V (I_C = 150mAdc, I_B = 15mAdc)	$V_{BE(sat)}$	—	1.3	Vdc

Table 5-1: Typical transistor data sheet (MPS-U02).

DYNAMIC CHARACTERISTICS				
Current Gain-Bandwidth Product (I_C = 20mAdc, V_{CE} = 20Vdc, f = 100MHz)	f_T	100	—	MHz
Output Capacitance (V_{CB} = 10 Vdc, I_E = 0, f = 100kHz)	C_{ob}	— —	— 20	pf

Table 5-1 *(continued)*

voltage rather than an experimenter/designer operating voltage. V_{CEO} usually tells you the collector emitter breakdown voltage with the base circuit open. Of course, you would not normally operate a transistor this way in a circuit. Nevertheless, you can use this specification. The 40 Vdc figure can be considered the *absolute maximum* voltage that you should use as a collector voltage during circuit tests.

A few points to keep in mind are:

1. In general, operate a transistor with a collector voltage less than the supply voltage (there are exceptions to this rule; for example, in certain rf circuits). But there are very few exceptions in low-frequency analog applications.
2. *Never repair a circuit board and then apply power if the source you are using is a higher voltage than the maximum voltage rating of the transistor, even with a resistor in the collector lead!*

The next two ratings are collector base voltage (V_{CB}), 60 Vdc max, and emitter base voltage (V_{EB}), 5.0 Vdc max. Again, these are test voltages rather than operating voltages. For example, in actual practice, you will usually have current flowing through the emitter base junction at all times. Therefore, the *real* voltage drop across this junction will be about 0.5 to 0.7 volts for a silicon transistor (about 0.2 to 0.4 volts for a germanium transistor).

In most practical breadboarding situations, you will have to select a bias (V_{EB} voltage) on the basis of what your input signal to the transistor is to be, rather than the spec sheet's given value. In any case, you should always consider any input signal that you wish to apply to the emitter base junction, in addition to what you set as a no-signal operating bias.

Collector Current

In Table 5-1, the collector current, continuous (I_C), is listed as 800 mA. It is important that you realize *collector current will increase with temperature.* This rating (and all the rest) is specified at 25°C (77°F). Notice that the next line, "Total Power Dissipation," is given as (@) T_A = 25°C.

Rule: *Do not operate any transistor at or near the maximum rating given in the maximum rating section of the spec sheet.*

Total Power Dissipation

In practical servicing problems, it is the *total power dissipation* (P_D) that are of major concern. Note that there are two given: T_A (operating) and T_C (collector). T_A, P_d is to be derated above 25° at 8.0 mW per degree Celsius, or 8.0 mW/°C. The other Total Power Dissipation, @ $T_C = 25$°C, should be derated above 25°C at 80 mW/°C. For example, assume that the collector operates at 40 V and 800 mAdc. This results in a power dissipation of 32 watts—far above the 10 watts specified for T_C at 25°C.

Operating and Storage Junction Temperature Range

How much power a transistor can dissipate is closely associated with the transistor's temperature range. As shown in Table 5-1, P_D must be derated above 25°C for each degree above the value given. The temperature range (T_j, T_{stg}) given is −55 to +150°C. In most cases, the high end (+150°C) is the most important. This is because, as a rule, collector current increases with temperature. That will, in turn, cause an additional rise in temperature, followed by an increase in collector current. Result: *thermal run-away!* End result: the transistor will burn out and you will have to replace it with another.

An external heat sink is required for maximum power output for all but the smallest semiconductor devices (power dissipation rating of less than 1 watt). If the area of the heat sink is insufficient to dissipate the heat generated within the semiconductor device (for example, an IC), in many instances an automatic thermal shutdown circuit will turn off the IC and prevent it from being damaged.

Thermal Characteristics

Notice that the thermal characteristics given in Table 5-1 are both listed as *Thermal Resistance*. Transistors designed for power applications (as is the MPS-U02 we are discussing), have these ratings specified to indicate power dissipation capability. Thermal resistance can be defined as the increase in temperature of the transistor *junction* with respect to either case (R_{Ojc}) or ambient temperature (R_{Oja}), or with respect to some other reference, divided by the power dissipation, or °C/W, as shown.

OFF Characteristics

The OFF characteristics given in Table 5-1 are strictly test values. For example, collector emitter breakdown voltage (BV_{CEO}) is the breakdown voltage (BV), with the collector (C) and emitter (E) connected to a voltage source and the base left open (O), or BV_{CEO}.

In this example, it is listed as 40 Bdc min. with the transistor non-biased. For this, we get *OFF characteristics*. Notice that each of the following includes a zero as the last symbol that indicated the transistor lead left open. For instance, BV_{CB0} indicates the emitter is open, and I_{CB0} also shows the same condition. However, as was explained before, test ratings such as these can be used for test purposes. The given values can be considered as the *absolute maximum* values you should apply to the device.

ON Characteristics

The symbol in this case, is h_{fe} (also known as *current transfer ratio*). You will see, by referring to Table 5-1, that the dc current gain for this transistor is given for three different conditions (the first being $I_C = 10$ mAdc, $V_{CE} = 10$ Vdc). The current transfer ratio of the transistor, under these conditions, is listed as an h_{fe} of a minimum of 50. Also, note that when I_C is equal to 500 mAdc @ $V_{CE} = 10$ Vdc the h_{fe} drops off to a minimum of 30. In other words, the transistor will not amplify as well at the larger collector current values. To repeat: *do not operate any transistor at, or near, its maximum current rating!* Although direct current characteristics are important for selecting proper bias circuits, they are primarily test values.

Dynamic Characteristics

Table 5-1 shows two transistor characteristics under the heading Dynamic Characteristics: Current Gain—bandwidth product, and Output Capacitance, both of which are important if you intend to use this transistor in high frequency circuits. Under the test condition given ($I_C = 70$ mAdc, $V_{CE} = 20$ Vdc, f $= 100$ MHz), the minimum current gain bandwidth of this transistor is 100 MHz. As a general rule, gain bandwidth product is the frequency (100 MHz, in this case) at which gain drops to a very low level. In some cases, this drop will be down to unity (0dB).

The other output capacitance (under the given test conditions) is a maximum of 20 pF. The lower this value (C_{ob}), the better. In general, capacitance will shunt the signal, resulting in a loss of amplification and a change of the expected circuit impedance.

SECTION 5.2: WORKING WITH CB/AMATEUR HIGH FREQUENCY TRANSISTORS

High frequency characteristics are especially important to the person working with two-way communications such as citizen band (CB) and amateur radio transistors. To properly match impedances, two components must be considered: output resistance versus frequency and output capacitance versus frequency. Usually, this information is provided on data sheets, in graphical form, over the frequency operating range of the transistor. Figure 5-1 shows a graph of these two transistor characteristics for the MRF475. This transistor is designed primarily for use in single sideband linear output application in CB and other rf equipment operating to 30 MHz.

The reason this information is presented by means of graphs or curves, rather than in tabular form, is that you need to know these resistance and reactance values (either capacitance or inductance) over a wide range of frequencies (1.5 to 30 MHz, in this example), not at some specific frequency (unless you happen to be testing a transistor for a specific frequency only).

There are two terms you will encounter, when reading rf transistor spec sheets, that may not be familiar: *carrier power* (P_C) and *peak envelope power* (PEP). Carrier power refers to the power available at the output terminals of the rf transistor (in its test circuit) when the transistor output terminals are connected to the normal load (usually 50 ohms). In this example, carrier power is typically 4.0 watts (continuous), with V_{CC} = 13.6 Vdc, and connected

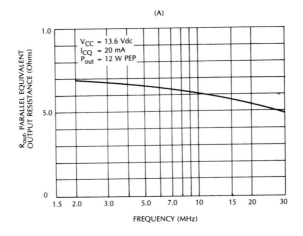

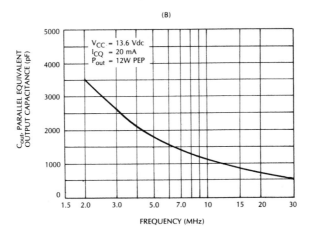

Figure 5-1: (A) Output resistance versus frequency. (B) Output capacitance versus frequency, for a MRF475. This device is a NPN silicon rf power transistor rated at 4.0 watts continuous wave (CW) @ 30 MHz.

to a 50 ohm load. Note that this is usually expressed as under *continuous* power output.

Peak envelope power is the *average power* supplied to some form of load (antenna, dummy load, etc.) by the transistor (with its test circuit), during one rf cycle at the highest crest of the modulation envelope, taken with the transistor and test circuit connected to its normal load. Again, this is 50 ohms in our example. However, the value for PEP is frequently derived under test conditions by doubling the original test setup power input, to simulate driver modulation.

When you divide PEP by P_C, subtract 1 and then multiply your answer by 100, this results in the percentage of *up-modulation*. In formula form, this is

$$\text{percentage up-modulation} = [(\text{PEP}/P_C) - 1] \times 100$$

This percentage of modulation is given in spec sheets at a certain P_C value (4.0 watts in the transistor spec sheets we are discussing), and, in this spec sheet, equals 100 percent. Percentage up-modulation and other tests can be measured and/or calculated by using the common emitter test circuit shown in Figure 5-2. See Table 5-2 for parts list (see next section for this figure and parts list).

If you have both an rf ammeter and an rf voltmeter, the power output is the product of the two readings regardless of the value of the load impedance, *provided the load you are using is a pure resistance.* This applies to a so-called *key-down* (operating) CW test circuit.

SECTION 5.3: MEASURING CARRIER OUTPUT POWER (SSB)

Measuring the output of a single-sideband (SSB) test circuit is a bit more complicated. Your best bet, if you don't have a regular rf test set available, is to use an oscilloscope.

For example, let's say that you are modulating a SSB test setup with a single tone, and carrier insertion is used. In this case, adjust both until you get all test patterns of equal heights, as shown on your scope. Turn the audio gain up on the SSB driver unit (it would be connected to the rf input shown in Figure 5-2), until the highest waveform is seen on your scope—*without distortion.*

Next, use an rf voltmeter and measure the peak voltage of the modulation pattern (the value from zero carrier to peak carrier). Now, convert this peak value to rms value by multiplying by 0.707, squaring, then divide by the value of your load (R_L). This will give you the peak envelope power of a SSB-excited test circuit such as the one shown in Figure 5-2. The formula for reading PEP on a SSB setup is

$$\text{PEP} = (E_{\text{peak}} \times 0.707)^2/R_L.$$

where R_L is the load resistance you use to terminate the test circuit.

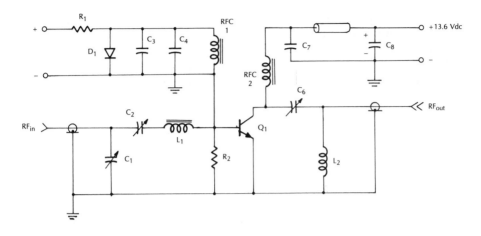

ADJUST FOR ICQ 20 mA

Figure 5-2: Common emitter test circuit for measuring carrier output power, transistor efficiency, power gain, etc. of the rf power transistor MRF475. See Table 5-2 for parts list.

C1, 2, 6	ARCO 466 Trimmer Capacitors.
C3	1000 μ F, 3.0 V_{dc} Electrolytic.
C4, 7	0.1 μ F, Disc Ceramics.
C8	100 μ F, 15 V_{dc} Electrolytic.
R1	10 ohm, 5.0 W Resistor.
R2	10 ohm, 1.0 W Resistor.
L1	2.2 μ H Molded Choke.
L2	4 turns #18 AWG wire, 1/2″ I.D., 5/16″ long.
RFC1	10 μ H Molded Choke.
RFC2	15 turns #20 AWG Wire on 5.6 k ohm 1.0 W Carbon Resistor
RFC3	5 Ferroxcube, #56–590–65/3B, Beads on #18 AWG Wire
D1	1N4997.
Q1	MRF475.

Table 5-2: Parts list for test circuit shown in Figure 5-2.

SECTION 5.4: GUIDE TO TESTING DRIVING RF OUTPUT TRANSISTORS

When a single transistor circuit is not capable of providing the required output power to another stage, an additional transistor amplifier is usually used. Another reason to use more than one amplifier stage is heat distribution. The use of two transistor amplifiers, rather than a single unit, splits the delivered power between the two stages.

As an example of how this can be done, let's use two transistors designed for 12.5-volt amplifier applications, in the 27 MHz CB radio band. The driver transistor could be a MRF8003. This transistor is rated at 0.5 watts, with a minimum gain of 10 dB, and an efficiency of 50 percent. A 27-MHz amplifier circuit schematic is shown in Figure 5-3. See Table 5-3 for parts list.

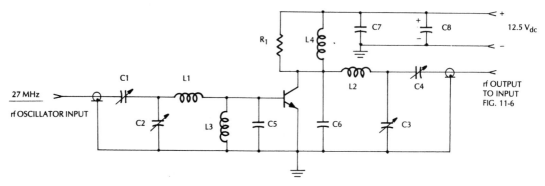

Figure 5-3: Schematic diagram for a 27-MHz transistor (MRF8003) amplifier driver stage. See Table 5-3 for parts list.

Parts List for Figure 5-3	
C_1, C_2, C_3, C_4	9.0 - 180 pF ARCO 463 or equiv.
C_5 _	25 pF UNDERWOOD.
C_6 _	100 pF UNDERWOOD.
C_7 _	1000 pF UNDERWOOD.
C_8 _	10 μF ELECTROLYTIC.
$L_1 L_2$ _	0.47 μH Molded Coil.
L_3 _	VK 200-20/4B RFC.
L_4 _	16 Turns No. 26 wire closewound on R_1.
R_1 _	390 ohms, 2 W.

Table 5-3: Parts list for circuit illustrated in Figure 5-3.

Notice that the transistor used in the schematic in Figure 5-3 is rated for only 0.5 watt. This transistor amplifier is suitable for use as a driver for the rf power transistor MRF8004, rated at 3.5 watts, 27 MHz. Figure 5-4 shows a schematic diagram for this amplifier. See Table 5-4 for parts list. It could be used as the final amplifier of a CB rig. Incidentally, no provisions for modulation are shown. Therefore it would be impossible to use these stages in a transmitter without including some form of modulating circuit. Of course, the system would also have to have a 27-MHz oscillator in conjunction with the driver and power amplifier circuits.

If you want to test the circuit shown, some form of output indicator should be used for loading (delivering the desired amount of power to each stage, and to the pure resistive load—dummy load—of the final stage). The output indicator may be an rf ammeter, rf voltmeter, or the forward position of an SWR meter. Where no modulation is applied, adjust the loading circuit (the variable capacitors) until you see no further increase at the output metering instrument. *Do not key (apply operating power) for more than 30 seconds at a time while making the loading adjustments.* Basically, the foregoing is also true for a SSB rf circuit—*unmodulated conditions.*

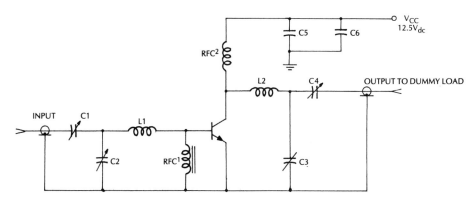

Figure 5-4: Schematic diagram for a 27-MHz transistor (MRF8004) amplifier, that can be used following the driver stage shown in Figure 5-3.

C_1, C_2 – 9.0 pF ARCO 463 or equiv.
C_3, C_4 – 5.0 - 80 pF RCO 462 or equiv.
C_5 – 0.02 μ F Ceramic disc.
C_6 – 0.1 μ F Ceramic disc.
RFC_1 – 4 Turns #30 Enameled Wire wound on Ferroxcube Bead Type 56-590-65/3B.
RFC_2 – 26 Turns #22 Enameled Wire (2 layers - 13 turns each$Z$) 1/4" inner diameter.
L_1 – 0.22 c Molded Choke.
L_2 – 0.68 c Molded Choke.

Table 5-4: Parts list for power amplifier circuit.

SECTION 5.5: GUIDE TO TESTING FM MOBILE COMMUNICATION TRANSISTORS

There are two general frequency bands utilized in FM mobile radios, for which Motorola manufactures transistors. There are 27- to 50-MHz (low-band FM transistors) where longer distance mobile communication is desirable, and 66- to 88-MHz midband transistors.

An example of a low-band FM transistor that can be used in rf amplifier circuits operating up to 50 MHz is the MRF402. Although this is a low output power transistor (1 watt @ 50 MHz), it does not require special mounting. It has a T0039 case rather than the typical rf packages shown in Figure 4-1 (see Section 4).

Figure 5-5 shows a test circuit and parts list that can be used to check the MRF402 transistor. The loading procedures are the same as explained in the last section. However, it bears repeating. *Do not apply operating power for more than 30 seconds at a time while adjusting the capacitors for maximum permissible power-out to the resistive load.*

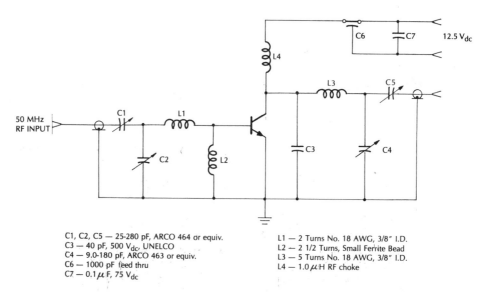

C1, C2, C5 — 25-280 pF, ARCO 464 or equiv.
C3 — 40 pF, 500 V$_{dc}$, UNELCO
C4 — 9.0-180 pF, ARCO 463 or equiv.
C6 — 1000 pF feed thru
C7 — 0.1 μF, 75 V$_{dc}$

L1 — 2 Turns No. 18 AWG, 3/8" I.D.
L2 — 2 1/2 Turns, Small Ferrite Bead
L3 — 5 Turns No. 18 AWG, 3/8" I.D.
L4 — 1.0 μH RF choke

Figure 5-5: Schematic diagram for a MRF402 test setup.

SECTION 5.6: MARINE RADIO TRANSISTORS/MODULES

The frequency range of these transistors and modules is from 156 to 1662 MHz. A transistor that is provided in the T-39 grounded emitter package, and will operate at 4 watts up to 175 MHz, is the MRF237. The grounded emitter TO-39 package is designed so that the transistor has high gain (12 dB minimum @ 175 MHz) and very good total power dissipation (8.0 watts @ T$_c$ of 25°C).

A VHF power amplifier module that operates in the same frequency range as the MRF237 transistor is the MHW613. Actually, this module operates at 150 to 174 MHz, with an output power of 30 watts. Figure 5-6 shows the package, with pins identified, for this rf power amplifier module. The MHW613 is designed for 12.5 volts VHF power amplifier applications in industrial and commercial equipment.

If you are working with equipment that utilizes a VHF power amplifier module such as the MHW613 or an equivalent, there are a few points to keep in mind:

1. Input impedance at pin 4, and output impedance at pin 1, is 50 ohms. In other words, that is the impedance you match when connecting external circuits to the module.
2. Pin 3 is a gain control pin that may be using either manual or automatic output level control.
3. The maximum rf input power to pin 4 is 500 mW. Do not exceed this rating if you are testing one of these modules. Normally, you should use about 300 mW with pin 2 set for 30 watts output power.
4. Do not exceed +16 Vdc on pins 2 and 3 during any test. You should use near 12.5 Vdc on these pins.
5. Due to the high gain of these units, decoupling networks are a must on both pins 2 and 3. These are basically low-pass filters and are generally included on the spec sheets.

Typically, marine radios are required to reduce their power-out to below 1 watt during in-harbor operations. You will find that most equipment designed around the MJW613 rf module achieves this by removing the dc voltage from P_2 (leaving P_2 open). Pin 3 will usually have 12.5 Vdc, and power input is normally achieved at 300 mV. Under these conditions, the module power-out will be between 0.3 watt to something under 1 watt.

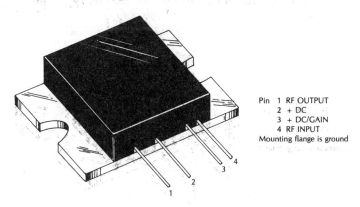

Pin 1 RF OUTPUT
  2 + DC
  3 + DC/GAIN
  4 RF INPUT
Mounting flange is ground

Figure 5-6: RF power amplifier module (30 W @ 150 − 174 MHz), MHW 613.

SECTION 6

Transmitter/Receiver Servicing

SECTION 6.1: PRACTICAL CB RADIO
TESTS AND MEASUREMENTS

With this section, and a little practice, you probably can accomplish a complete system "checkout" in not much more time than it takes to read each individual test or measurement needed. You'll find this section explains a variety of CB tests and measurements that can be performed by any electronics technician.

While it's true that many electronics technicians/CB operators do not have extensive electronics shop facilities, this should not deter them from enjoying the benefits of the many tests which require no more than simple, homemade test gear (in fact, in many tests, no test equipment at all is needed) found in the following pages. CB system checkout—from both a technical and cost point-of-view—is simple. You can do it yourself.

Although you can service CB equipment with a nonradiating dummy antenna connected to a transceiver, it is also *important* you realize that if any *internal adjustments* are made, the set must be checked out by a person holding a valid FCC license before it is operated using a radiating antenna. Follow these rules and you can't go wrong:

1. Do not attempt to operate any transceiver until a suitable dummy antenna is connected to the set (assuming you have it in a shop for testing).
2. Don't operate the CB unit for extended lengths of time until you have checked the standing wave ratio (SWR). You can tolerate an SWR up to 3 to 1 for short periods of time. However, the output power will be greatly reduced, and it's possible to damage the unit with an SWR greater than 3 to 1.
3. Don't test the system on the air without complying with FCC rules and regulations. FCC regulations strictly limit the amount of "on-the-air" testing that may be done and, as previously explained, after anything is done to a set that affects frequency, modulation, or power

(in other words, any internal adjustments or repairs), your work must be checked by a technician holding an appropriate FCC license.

Timesaving Antenna Test

Test Equipment:

12-volt car lamp (incidentally, a number 1815 or 1487 12-volts pilot lamp will also work) and screwdriver or jumper lead.

Test Setup:

See Figure 6-1.

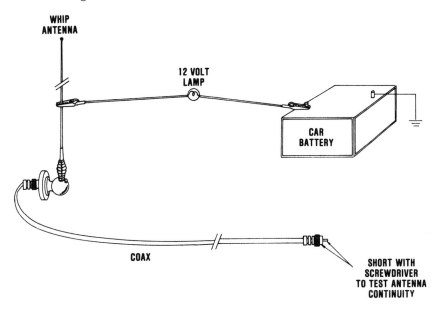

Figure 6-1: Test setup for a CB transmitter antenna open or shorted circuit test.

Comments:

Sometimes it's very difficult to get to antenna feed lines. One of the easiest ways I've found to check an inacessible coax transmission line without instruments is this simple procedure:

Procedure:

Step 1. Disconnect the coax cable going to the antenna from the CB transceiver by unplugging the cable's coaxial connector that goes into the antenna jack on the back of the transceiver. *Under no circumstances should you trim the coax without consulting the manufacturer's instructions.* In many systems, the transmission line is "tuned" for a specific length to present a proper impedance.

Step 2. Connect one alligator clip to the antenna whip and the other to the car battery's "hot" terminal, as shown in Figure 6-1.

Step 3. Use a screwdriver or clip lead and short between the coax connector center pin and outside shell. If the lamp does not glow, the coax line, one of its connectors, or base connection is open.

Step 4. It's also possible that the lamp will light before you short the coax plug; in other words, just as soon as you connect the bulb, as shown in Figure 6-1. In this case, you may have a short in the antenna system. It could be the coax cable, antenna, or one of the connectors is shorted (another reason is explained in Step 5). I always check the coax connectors first, because they have proven to be more troublesome than the antenna or cable.

Step 5. The other reason your lamp may light is that the antenna is a shunt-fed, base-loaded type. This type of system is easy to spot because it usually has a loading coil clearly visible in the whip. You'll have an operating lamp (with or without the coax plug shorted) with this system.

AM Modulation Measurements

Quick Modulation Check

Test Equipment:
Lamp-type dummy load (see dummy loads in this section)

Test Setup:
Connect a lamp-type dummy load to the transceiver antenna jack.

Procedure:
Step 1. Energize the transmitter and note the brilliance of the lamp as you speak into the mike.

Step 2. If the transceiver is modulating the rf carrier, you should see the lamp flicker. It should become brighter and dimmer, depending on the loudness of your voice (in AM sets). With a little bit of practice, you can learn to make an estimate of the percent of modulation.

Voltmeter Method

Test Equipment:
RF voltmeter, dummy antenna, audio-signal generator

Test Setup:
See Figure 6-2.

Comments:
In most cases, you'll find that the transmitter modulation is okay if the receiver audio is operating properly, because both use the same stages. However, if you do have trouble, using this method of checking modulation is

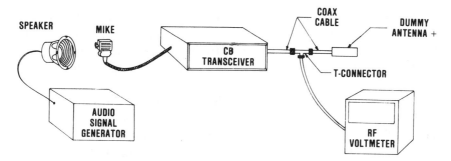

Figure 6-2: RF voltmeter connected to a CB transceiver and dummy load through a T-connector.

one way to get around the problem of not having an oscilloscope. Furthermore, it's even possible you may not need any additional equipment if you already have an rf ammeter in the system. In either case, with an rf voltmeter or rf ammeter, all you have to do is note the meter readings during modulation and refer to Table 6-1 to determine the percent of modulation.

ANTENNA CURRENT INCREASE WITH MODULATION	PERCENT OF MODULATION
22.5%	100%
18.5%	90%
15.5%	80%
11.5%	70%
8.6%	60%
6%	50%
4%	40%
2.2%	30%
1%	20%
0.25%	10%

Table 6-1: RF voltage (or current) increase versus percent of modulation.

Procedure:

Step 1. Connect the equipment as shown in Figure 6-2.

Step 2. Modulate the transmitter by feeding 1,000 Hz acoustically into the mike (highest permitted modulation frequency is 3,000 Hz). Increase and decrease the output amplitude control on the signal generator.

Step 3. Observe the voltmeter (or ammeter) reading as you vary the signal generator output level control.

Step 4. Refer to Table 6-1 to determine the percentage of modulation.

Step 5. If the set overmodulates (more than 22.5 percent voltage increase) on loud tones—a common trouble—check the AMC adjustment. *Overmodulation may cause severe television interference during operation of the CB rig.*

Step 6. Weak modulation is an indication of mike or transistor troubles. Disconnect the mike cable and inject your 1,000 Hz test tone (set at about 35 mV rms for 100 percent modulation) into the mike input jack. If you can't get 22.5 percent increase in your voltage (or current) reading between no-tone (zero voltage input and a loud tone (about 35 mV rms input), it's a mike-amplifier or AMC defect.

Oscilloscope Method—Sine-Wave Pattern

Test Equipment:

General purpose oscilloscope, radio receiver that tunes to the CB frequency of interest, small rf coupling capacitor, and an audio-signal generator (optional)

Test Setup:

Connect the scope's vertical input terminal to the receiver's last IF amplifier transistor collector lead, using a dc blocking capacitor in series with the scope's "hot" lead, as shown in Figure 6-3.

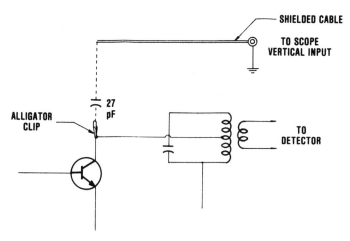

Figure 6-3: Scope connection using a radio receiver to check for overmodulation in a CB transmitter.

Comments:

To actually "see" the rf carrier on an oscilloscope, requires a scope with a vertical input bandwidth up in the 30- or 40-MHz bandwidth region. Of course, this puts the scope in a higher price range. However, there is a way out of the dilemma. Use a down-converter. In other words, pass the modulated carrier through a receiver mixer and look at the lower intermediate frequency at the receiver's last IF amplifier. *Note:* It is all but impossible (except for a very rough estimate) to measure percent of modulation using this procedure. However, what is most important, you can quickly detect *over-modulation*, which is what makes the FCC unhappy. Figure 6-4 illustrates some of the patterns you'll see on your scope.

Procedure:

Step 1. Connect the oscilloscope vertical input test lead to the receiver's last IF amplifier collector lead, through an rf coupling capacitor.

Step 2. If you're testing the transceiver in-shop, place a dummy antenna on the antenna terminals *before energizing the set.* Don't worry about picking up the signal on the receiver being used for a down-converter because there should be ample radiation anywhere in your shop. If not, connect any piece of wire (for example, an unused test lead) to the receiver as an external antenna.

Step 3. Key the transmitter and apply sinusoidal modulation. Modulation can be done by acoustical means (through the mike), or with an audio-signal generator connected to the mike input of the transceiver under test. If you use a signal generator at the mike input, set the frequency for 1,000 Hz, with an output level of about 35 mV.

Step 4. Check the pattern on the scope. The most important ones you'll see are shown in Figure 6-4.

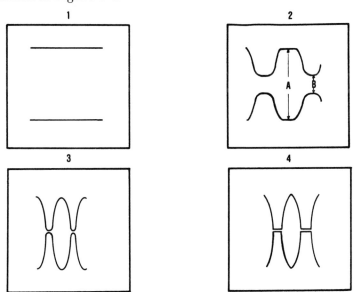

Figure 6-4: Approximate modulation patterns you'll see on an oscilloscope under various conditions. (1) RF carrier, no modulation, (2) less than 100% modulation, (3) 100 percent modulation, (4) overmodulation.

Step 5. A rough approximation of the percent of modulation can be calculated by using the formula:

$$\text{percent of modulation} = [(A - B)/(A + B)] \times 100$$

where A and B are as shown in Figure 6-4 (2). What you need is to be sure that the set is capable of more than 70 percent but not over 100 percent, the modulation is linear (appears equal on both top and bottom), and is sinusoidal.

Dummy Loads, Construction and Use of

Lamp-Type

Test Equipment:
   Number 47 lamp and phono plug

Test Setup:
   See Figure 6-5.

Comments:
   Do not make internal adjustments on any CB transceiver when it is connected to a radiating antenna. Always use a dummy antenna. A simple one can be constructed using a pilot lamp and phono plug and may be used to terminate the CB rig during testing.

Procedure:
   Step 1: Solder the circuit together, as shown in Figure 6-5.

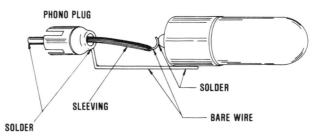

Figure 6-5: Construction details for a lamp-type dummy antenna.

Step 2. Plug the completed dummy antenna into the transceiver antenna jack, for testing. The lamp acts as a load of approximately 50 ohms. It also will provide you with a means of checking relative power output and modulation.

Coaxial Types

Test Equipment:
   Coaxial plug PL-259 (sometimes called a *VHF plug*) and a 1/2 watt composition resistor

Comments:

The reason for the copper disk shown in Figure 6-6 is to reduce the lead inductance and provide shielding.

Construction Details:

Step 1. Select the correct value composition resistor for the make (or makes) CB set to be serviced (usually, approximately 50 ohms).

Step 2. Mount and solder the resistor in the coaxial plug as shown in the cut-away in Figure 6-6.

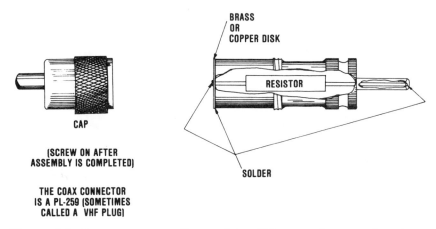

Figure 6-6: Dummy antenna for servicing CB transceivers, made using a composition resistor and coaxial plug.

Audio Load Resistor Power Measurement

Test Equipment:

Audio load resistor and rms reading voltmeter

Test Setup:

Substitute the audio load resistor for the speaker.

Comments:

In most cases, it is best to make receiver tests with the receiver terminated by a resistor, with resistance equal to the speaker impedance. The resistor should be rated at least 50 percent over the expected power dissipation.

Procedure:

Step 1. Refer to the service notes to find the rated power output of the equipment under test. This is the power that should be delivered to the audio load resistor.

Step 2. Measure the voltage across the load resistor with the rms reading voltmeter and use the indicated value to convert power to equivalent voltage, by using the formula $E = \sqrt{PR}$, where E is the rms voltage, P is the power listed in the service notes, and R is the load resistance in ohms.

Step 3. If you service quite a few two-way radio sets, it's well worth your time to use a calculator and work up a table, or graph, to convert power to rms voltage for the common audio loads (say 3.2 and 8 ohms). A sample table for an 8-ohm load resistor is shown in Table 6-2.

POWER WATTS	VOLTS (rms)
0.5	2
1	2.8
2	4
3	4.8
4	5.6
5	6.3

Table 6-2: With an 8-ohm resistor, this table can be used to convert required power readings to voltage readings.

Simple CB Radio Power Input System Test

Test Equipment:
    VOM

Test Setup:
    Disconnect the CB set from the vehicle's electrical system or, if it's a base station operating off the ac power lines, pull the wall plug.

Procedure:
    Step 1. Set your VOM to read on its lowest resistance range (for example, R × 1 or R × 10).

    Step 2. With the set's ON-OFF switch turned at ON, measure the resistance across the CB unit power input leads.

    Step 3. If your resistance reading is near zero, the CB transceiver power input system probably is operating correctly.

    Step 4. If Step 3 checks out okay, your next step is to check the vehicle's electrical system. In the case of a base station, check for about 120 Vac at the wall plug. Incidentally, a quick way to check an automotive system is to connect the CB set directly to the car battery .

Step 5. On the other hand, if you read a very high resistance (or infinite resistance), there is probably trouble in the power cable, plug, fuse, or the power circuit wiring inside the set.

Mobile Transceiver System Bench Testing

Test Equipment:

Power supply/battery eliminator, ammeter (at least 200 mA), voltmeter, and dummy antenna

Test Setup:

See **Procedure.**

Comments:

Inexpensive test equipment can be used for rough performance tests. However, the cables and controls used should duplicate the conditions of normal equipment use, or your measurements and test will be further invalidated. For example, to simulate vehicular installations, the transceiver should be connected to the ungrounded battery eliminator terminal with about 7 feet of No. 6 copper conductor. If you don't do this, the voltage drop on the vehicle cable contributes to the error and invalidates, to a greater degree, the measurement data obtained at the bench. This is most important when performing tests on trunk-mounted units.

Procedure:

Step 1. Connect the transceiver to the dc voltage workbench power supply. Watch polarity! Make sure the ground lead is not inadvertently connected to the "hot" power supply terminal.

Step 2. Place an rf dummy antenna on the transceiver antenna plug. You'll find construction details for a dummy antenna in this section.

Step 3. Place your ammeter between the bench power supply and transceiver power input cable. You can expect between 100 mA and 200 mA of current in a properly operating receiver. Therefore, set the VOM range switch to a range of 200 or above.

Step 4. Turn on the bench supply and set the output at 13.6 V, ±0.5 V.

Step 5. Note the receiver idling current (no signal being received) with the squelch open and the volume set at maximum. As was mentioned in Step 3, you'll probably measure a current between 100 and 200 mA, depending on the set. If your reading is above these values, it is possible that you have a shorted protective diode in the receiver. If you have to replace a diode, typically you'll find that they are rated at 1.5 amps with a peak inverse voltage (PIV) rating above 300 Vdc.

Easy-to-Make Transmitter Carrier Frequency Checks

Radio Receiver Method

Test Equipment:
 Multiple-band calibrated radio receiver, dc voltmeter (the voltmeter is not needed if the receiver has an S-meter), and an rf dummy load (construction details for a dummy antenna are given in this section).

Test Setup:
 Place a dummy antenna on the CB transmitter and loosely couple the transmitter to the receiver.

Comments:
 If you don't have a suitably calibrated radio receiver with a built-in S-meter, it's possible to connect a dc voltmeter between the AGC bus line and ground. The voltmeter will act as a tuning indicator (S-meter). Or, you can contact a local amateur radio operator. It's quite probable that he will help you, assuming that you only want to check your own CB rig. Many of the radio receivers used by amateur radio operators have a self-contained S-meter and crystal-controlled oscillator that provide checkpoints for receiver calibration. However, the simple frequency check described here does not meet the FCC technical standards, as defined for FCC record purposes. It is important you realize that a transceiver *must remain within 0.005 percent tolerance* (approximately 1850 Hz), to comply with FCC rules and regulations.

 When using an S-meter, be careful how you interpret the readings. For example, depending upon the rig, an S-meter may be calibrated so that one S unit is equal to 6 dB. Some S-meters, however, are calibrated at only 3dB per S unit, and others at 3 or 4 at the low end and 6 or 7 at the top of the scale. Just remember, if you use different S-meters, your readings may vary considerably.

 Another problem encountered with the S-meters, is their inability to measure high strength inputs. When checking a powerful signal, some S-meters will appear erratic in operation and even read lower on the scale, as the signal level is increased.

Procedure:
 Step 1. Place a suitable dummy antenna (usually 50 ohms) on the CB transceiver under test. Use a nominal supply voltage (13.6 Vdc for mobiles, 117 Vac for base stations), and conduct the test at room temperature (70°F).

 Step 2. Energize the receiver and CB transceiver and let them warm up for 30 minutes or so. It isn't necessary to energize the transmitter during the warm-up.

 Step 3. If the receiver has a self-contained calibrator (for best results, it should have one), calibrate the receiver.

 Step 4. Energize the CB transmitter and tune the receiver for peak deflection on the S-meter, using loose coupling between the two (as much distance between the two pieces of equipment as practical).

Step 5. Read the CB transmitter frequency directly from the receiver tuning dial.

Field-Strength Meter Method

Test Equipment:

Frequency calibrated field-strength meter and dummy antenna.

Test Setup:

Place a suitable dummy antenna (usually 50 ohms) on the CB transceiver and loosely couple the transceiver to the field-strength meter. Use a nominal supply voltage (13.6 Vdc for mobile, 117 Vac for base stations), and conduct the test at room temperature (70°F).

Procedure:

Step 1. Set the field-strength meter and transmitter to the frequency to be measured.

Step 2. Turn the transmitter on and adjust the field-strength meter for peak deflection. If you're using antenna coupling (some field-strength meters use more than one method of coupling; i.e., capacitance, cable, etc.), use loose coupling. This can be achieved by varying the distance between the equipment.

Step 3. When you have the field-strength meter tuned exactly to the transmitter frequency (indicated by peak deflection), read the frequency off the field-strength meter tuning dial.

Transmitter Power Output Measurement

Measurement Equipment:

RF wattmeter (see **Comments**)

Measurement Setup:

See Figure 6-7.

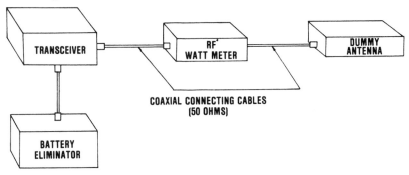

Figure 6-7: Test setup for measuring a CB transmitter's output power.

Comments:

When servicing CB transmitters, it is better to use one of the bidirectional power meters that are designed to meet the needs of a CB serviceman. For example, they are capable of measuring low forward power (20 mW to 30 mW), and will also measure the reflected power, whereas many power meters won't budge off zero scale reading at 4 watts forward power.

You'll also find that you may need three additional pieces of test gear, depending on the power meter and test setup you have to use: (1) a dummy antenna (usually 50 ohms), (2) a battery eliminator or other type of dc power supply if you're testing on the bench, and, (3) an N-type coax connector to PL-259 adaptor in your coax cable connections.

Procedure:

Step 1. Connect the equipment, as shown in Figure 6-7. If the power meter has an internal dummy load, you will not need the external one shown.

Step 2. Measure the power output with the rf power meter (if you're servicing single-sideband (SSB) equipment, skip Steps 3 and 4).

Step 3. If the output power is below 3 watts (unmodulated), the transmitter is in need of repair or alignment. Try peaking the tuned circuits on the driver and final stages.

Step 4. If the output is over 4 watts (unmodulated), *under no circumstances* should the transmitter be connected to a radiating antenna because this is a direct violation of FCC regulations. In this case, measure the dc supply voltage. You should not have more than 14 Vdc. If it's higher, adjust it back to 13.6 volts.

Step 5. When checking a CB transmitter (SSB), you should use a peak-reading rf wattmeter and measure "peak envelope power" (PEP). It is important to remember that PEP output is modulated power and, therefore, much greater than unmodulated power. However, what you want to read is 12 watts PEP on SSB. Anything over this is an FCC violation.

Crystal Test

Test Equipment:
Two operating CB sets

Test Setup:
See **Procedure.**

Comments:

Many inexpensive CB sets are designed to use a frequency synthesizer; the most popular being a 6-4-4 type. Figure 6-8 shows a block diagram of a typical synthesizer, listing the crystal frequencies, and Table 6-3 lists the crystal combinations.

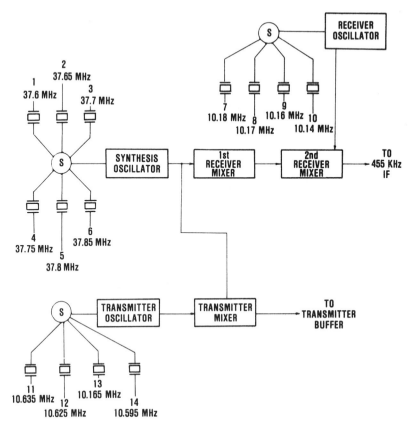

Figure 6-8: Block diagram of the most frequently used synthesizer in low-cost CB units.

Procedure:

Step 1. Check the performance of two CB rigs. This can be a base station and mobile, two hand-held sets, or any combination of two similar CB transceivers.

Step 2. After you're sure that the two sets are operating properly, simply remove one of the internal crystals and replace it with the crystal of unknown quality.

Step 3. Again, check and see if you can establish communications between the two. If you can, the crystal may be okay. To make a positive test, measure the transmitter, or receiver, performance and *above all, measure the set's frequency before placing it on the air.* Failure to do this, can result in an FCC citation for off-frequency operation. Be sure and measure the rf output frequency on every channel (the easiest way to do this is to use a frequency counter). All should be within 1 kHz of the assigned carrier frequency.

CHAN-NEL	CRYSTAL		CHAN-NEL	CRYSTAL		
1	7	11	13	4	7	11
2	8	12	14	4	8	12
3	9	13	15	4	9	13
4	10	14	16	4	10	14
5	7	11	17	5	7	11
6	8	12	18	5	8	12
7	9	13	19	5	9	13
8	10	14	20	5	10	14
9	7	11	21	6	7	11
10	8	12	22	6	8	12
11	9	13	23	6	10	14
12	10	14				

Table 6-3: Crystal combinations used in a 6-4-4 synthesizer.

Receiver Crystal Check

Test Equipment:
Properly operating CB transceiver, frequency-calibrated variable-frequency rf oscillator

Test Setup:
See Figure 6-9.

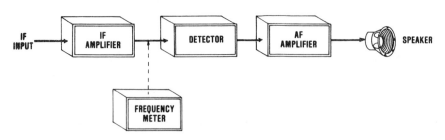

Figure 6-9: Test setup for checking receiver crystals.

Comments:

Typically, a so-called *low-cost precision rf signal generator's* calibration accuracy is ±3 percent, which is not good enough to permit use of one of these signal generators as a frequency standard.

Procedure:

Step 1. Open the CB receiver under test and loosely couple a frequency meter to the last IF, as shown in Figure 6-9.

Step 2. Energize the signal generator, CB receiver under test, and a separate CB transmitter, all tuned to the same frequency.

Step 3. Tune the signal generator until you have zero beat.

Step 4. Note the frequency reading on the frequency generator and compare it to the receiver manufacturer's published IF frequency (usually 455 kHz or 1650 kHz).

Step 5. In 6-4-4 synthesizers, four crystals commonly are used in the receiver oscillator. If the measured IF differs widely from channel to channel, it's an indication that the receiver crystals may not be correctly matched. A 6-4-2 unit (not as popular as a 6-4-4) uses one transmit crystal and one receive crystal. Channels that are on frequency for the transmit mode are accurate for receive, if the receiver oscillator is operating properly.

Receiver Audio Stage Test

Test Equipment:

Audio-signal generator and a good-quality ac voltmeter

Test Setup:

See **Procedure.**

Comments:

In most cases, you'll find the receiver audio stages are working properly, if the transmitter modulation is found to be correct, because both use the same stages.

Procedure:

Step 1. Set the audio-signal generator to operate at 1,000 Hz and connect it to the output side of the AM detector diode.

Step 2. With the receiver under test, and the signal generator turned on (and connected as explained in Step 1), measure the output of the signal generator and set it to about 0.2 volts rms. You can use a scope to make this setting (about 0.5 volts peak-to-peak).

Step 3. Measure the rms voltage across the speaker terminals.

Step 4. Next, use the formula $P = E^2/R$ to calculate the power delivered to the speaker. You'll probably have slightly more than 3 watts. For example, with an 8 ohm speaker and a 5 volt rms voltmeter reading:

$$P = E^2/R = 5^2/8 = 3.125 \text{ watts}$$

Squelch Test

Test Equipment:
RF signal generator that can produce a modulated rf signal level of about 1 μV

Test Setup:
See **Procedure.**

Procedure:
Step 1. With no modulated rf signal, set the squelch control so that it just cuts off all noise.

Step 2. Couple the rf signal generator to the transceiver antenna and apply a modulated signal of about 1 μV. If the squelch is operating properly, the modulated signal should cause the receiver to again start to operate.

Step 3. The squelch control now should be set at about 1/3 rotation, to cut off audio with no signal applied. If it is not, loosen the knob set screw and set it to this position.

Automatic Gain Control (AGC) Test

Test Equipment:
RF signal generator that can produce a modulated rf signal from 5 μV to 100,000 μV

Test Setup:
See **Procedure.**

Procedure:
Step 1. Set the signal generator to produce a modulated rf output signal in the channel the receiver under test is tuned to, and adjust its output level for 10 μV.

Step 2. Couple the generator's output signal to the CB transceiver antenna.

Step 3. Connect an ac voltmeter across the speaker terminals and measure the audio voltage (don't disconnect your voltmeter).

Step 4. Increase the signal generator rf output level to 10,000 μV.

Step 5. Again, note the voltmeter reading. Now, the readings in Step 3 and Step 5 (this Step) should not show an increase of more than 10 percent, if the AGC is operating properly.

Phase-Locked Loop (PLL) Test

Test Equipment:
VOM and good quality (well-filtered, well-regulated) dc power supply and a frequency counter.

Test Setup:
See **Procedure.**

Comments:

The following procedure is designed to test most typical, low-cost, 23-channel CB transceivers that incorporate a PLL in their design.

Procedure:

Step 1. Measure the dc voltage on the input of the voltage controlled oscillator (VCO). Refer to the set's schematic for the correct voltage. Typically, the voltage will be in the 1- to 2-volt range. If this voltage is off more than 4 or 5 tenths of a volt, it can cause the VCO to operate off frequency and, in severe cases, kill the oscillator completely.

Step 2. Connect a dc power supply to the VCO dc control voltage line (between the phase detector and VCO). Also, connect your VOM to the same point (many sets have a test post in this line).

Step 3. Using the VOM, set the power supply output voltage to the level indicated on the set's schematic. Again, it will probably be between 1 and 2 volts.

Step 4. Next, connect a frequency counter to the VCO output.

Step 5. Refer to the schematic. Your counter should read within the

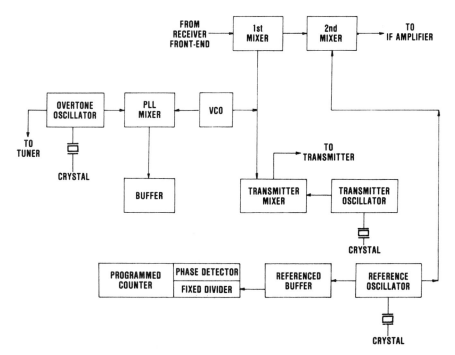

Figure 6-10: Block diagram of a typical 23-channel PLL system.

limits set by the manufacturer. You'll find these PLLs should operate near several different center frequencies. For example, sets using PLL-01A ICs operate the VCO at a frequency of 21 MHz. Sets using a separate reference oscillator, operate at 6.4 MHz, and the PLL-02A (and most others) operates at about 38 MHz, using a 10.24 MHz frequency reference.

If you can't get the VCO to operate at the proper frequency by adjusting the power supply to the voltage shown on the schematic, it's an indication that the VCO needs adjustment or replacement.

Step 6. Next, check the loop stage-by-stage, referring to the schematic and/or service manual. However, servicing is made much easier if you can get the system to operate at any frequency, using the dc power supply and adjusting its output to whatever voltage is needed to start the system operating. Before you start troubleshooting, try a voltage slightly higher and slightly lower. It may start the system operating. Once you get it going, all that's left to do is refer to the schematic for the correct voltages to locate the stage at fault. A block diagram of a typical PLL is shown in Figure 6-10.

SECTION 6.2: PRACTICAL GUIDE TO MEDIUM POWER TRANSMITTER TESTS AND MEASUREMENTS

We are all interested in just how well modulated our transmitter is, and it's common knowledge that the closer a transmitter is to having 100 percent modulation, the better the received signal will be. But overmodulation can cause interfering signals (splatter) and, possibly, bring in an FCC "ticket" as well. This section includes a simple scope circuit you can build which will permit you to skillfully monitor the modulation level of any AM transmitter—old or new—even if you don't own an oscilloscope.

The most important transmitter checks are power-out, modulation, and frequency. You'll find all these explained in the following pages. This section will show you how to expertly test a radio transmitter—AM, FM, or SSB using only simple and fast steps. You'll also find how to measure the performance of your transmitter, how to get the most out of it with the highest quality of performance and, in most cases, with inexpensive test equipment.

Included are easy-to-build coupling devices that can be used with all kinds of test gear, with the construction details clearly shown. There are oscilloscope patterns for checking SSB, AM and CW transmitters. Furthermore, in every test and measurement, you get complete step-by-step instructions that can save hours of time and effort when testing all kinds of transmitters.

Simplified Measurements of Unmodulated AM RF Power Out

Test Equipment:
Voltmeter with rf probe and a suitable dummy antenna (see **Comments**)

Test Setup:
    See **Procedure.**

Comments:
    Quite a few of the power measuring instruments on today's market are *directional* wattmeters; for example, the power/SWR tester sold by Radio Shack and the similar wattmeter/SWR bridge sold by Heathkit. While it's true that directional wattmeters are the best way to check antenna system performance, they aren't necessary for conducting standard medium-power transmitter power measurements. All you really need is a voltmeter with an rf probe and a dummy antenna. Figure 6-11 shows a schematic of a commercially built, 50 ohm termination.

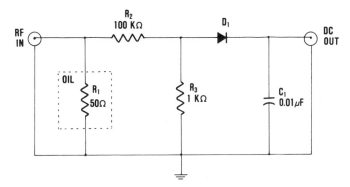

Figure 6-11: Schematic of a dummy antenna. It handles up to 1kW with VSWRs of less than 1.5:1.

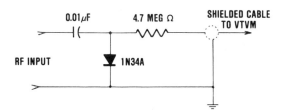

Figure 6-12: Schematic of an rf probe that can be used with a dc voltmeter to measure the peak value of an rf signal.

    If you don't have an rf probe, it's easy to construct one. Probably, the simplest and best-known way to build an rf probe is shown in Figure 6-12. The only thing that you'll have to be careful of is the voltage rating of the diode. The one shown is rated for a maximum peak reverse voltage of 75 V and a maximum forward voltage of 50 V, at a maximum forward current of 5 mA. To determine actual rf power out of the transmitter, you'll have to measure the voltage across the rf input to the dummy load. The following procedure explains how to do this.

When using the probe and dc voltmeter, you'll find that accuracy (about 10 percent, under ideal conditions) is best near the top end of the scale. To help improve your accuracy, use as few connectors as practical in your connecting cables, and all cables should be assembled as carefully as possible. Be sure to remember when working with radio frequencies that every transmission line connector is a potential source of impedance discontinuity, which can cause an increased SWR and power loss.

Procedure:

Step 1. Terminate the transmitter with a dummy antenna. Whatever you use as a dummy antenna, remember, this load must dissipate the power output of the transmitter under test for long periods, without physical damage or impedance changes.

Step 2. If the transmitter does not have a built-in, stable dc supply voltage, connect an external power supply to the unit.

Step 3. Connect the voltmeter's demodulation probe across the transmission line feeding the dummy antenna.

Step 4. Adjust (or re-check) transmitter tuning, loading, coupling and/or power adjust control settings.

Step 5. Read the voltmeter dc scale. Then use the formula:

$$\text{Power} = \text{voltage}^2/50 \text{ ohms}$$

Warning: If you use the circuit shown in Figure 6-12 with your VTVM, it will read the *peak value* of voltage on the dc scale. To convert to rms voltage, use the formula:

$$\text{rms voltage} = 0.707 \times \text{peak voltage}$$

Simple RF Current Measurement

Test Equipment:

VOM, demodulator probe, 10-ohm carbon resistor (10 watts or more), and/or a dummy antenna (usually 50 ohms)

Test Setup:

See **Procedure.**

Comments:

If you don't have a resistor capable of handling the power output of the transmitter under test, simply use resistors in a series-parallel combination to achieve the required power rating and current impedance.

Procedure:

Step 1. Place a 10-ohm, 10-watt carbon resistor in series with the ground side of the transmission line you're going to use to feed the dummy antenna.

The reason for using the ground side of the transmission line is to prevent accidental shorts.

Step 2. Connect the transmission line and resistor network to the transmitter under test.

Step 3. Connect the demodulator probe across the resistor. Set the voltmeter to a low dc scale (about 10 volts midscale, in this case).

Step 4. Energize the system and measure the voltage, then convert your voltage reading to current, using the formula I = E/R. For example, let's say you read about 14 Vdc. Now, this is peak voltage (using a peak-reading demodulator probe), so 0.707 × 14 = approximately 10 volts. Therefore, 10/10 = 1 ampere. Next, knowing the current, calculate the power.

$$P = I^2R = 1^2 \times 10 = 10 \text{ watts}$$

To measure the higher power, use a higher power resistor and measure the voltage across the dummy antenna. As an instance, into a dummy load (50 ohms) with a voltage reading of 100 volts rms, the current is 2 amperes. The power comes out to be 200 watts.

AM Modulation Monitor and Measurement Procedure

Test Equipment:
"Home-brew" circuit shown in Figure 6-13

Test Setup:
See **Procedure.**

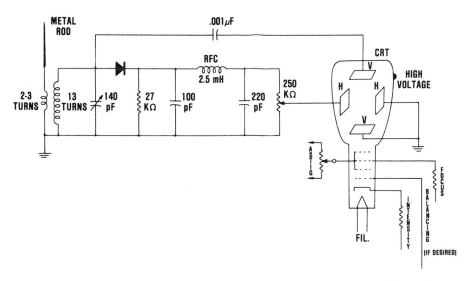

Figure 6-13: Circuit diagram for an AM modulation monitor that will produce a trapezoidal modulation pattern during use.

Comments:

If you don't have an oscilloscope, you can build a suitable power supply for almost any electrostatic CRT that, when connected to the circuit shown in Figure 6-13, will function as a modulation monitor. Other systems of measuring modulation are described in Section 6.1 and in Section 5.

The coil (L_1) and capacitor (C_1) you use in the tuned input circuit are dependent on the frequency band the transmitter works in. The rest of the circuit is simply a diode detector and filter that couples the demodulated audio signal to the CRT horizontal deflection plates, plus one line and capacitor to the vertical plates. If the output of the transmitter is strong enough, you can get by without the antenna, its coil, and capacitor C_1. Almost any general purpose or high frequency diode (such as a 1N34A or 1N6) can be used in the circuit.

Procedure:

Step 1. Turn on the CRT power supply. You'll see a spot on the screen when it warms up. Incidentally, don't leave the spot on too long before going to the next step because it's possible that you might burn the CRT face plate.

Step 2. Key the transmitter and move the monitor antenna (or whatever you're using to couple the monitor to the transmitter) to several positions until the CRT spot changes to a vertical line. Adjust coupling until a vertical line fills about two-thirds of the CRT screen.

Step 3. Speak into the microphone and you should see the line change into a trapezoid similar to one of the patterns shown in Figure 6-14. Use the 250 k-ohm pot to adjust the width of the pattern.

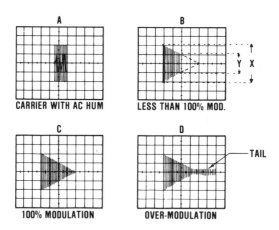

Figure 6-14: Trapezoidal patterns that may be seen while observing the modulation monitor described in the text.

If you see the pattern shown in (A), you have ac hum problems (could be happening inside the transmitter or being picked up from the lights, etc. in the room), (B) is normal but showing less than 100 percent modulation (see the next step), and (C) is within legal limits and what you should see at 100 percent modulation. You want to be able to produce this pattern, or come very close to it, for best reception at the receiver end. (D) shows a "tail," which means the transmitter is being overmodulated and, if you radiate a signal under these conditions, it's sure to cause you much trouble with the FCC, your neighbors (it can cause television interference), and other radio operators, because of splatter.

Step 4. For this step, use Figure 6-14 (B) and the two points on the waveform marked (X) and (Y). To calculate the approximate percent of modulation, inject an audio signal into the transmitter microphone input and then measure the height of the two points. After you make the measurements, simply use the percent of modulation formula:

$$\text{percent of modulation} = [(X - Y)/(X + Y)] \times 100$$

As an example, let's assume you measure 2.5 inches for the larger vertical (X) and 1/2 inch on the smaller vertical (Y). Substituting the measurement values for the letters in the formula, we get,

$$\text{percent of modulation} = [(2.5 - 0.5)/(2.5 + 0.5)] \times 100 = 66\%$$
$$\text{modulation}$$

CW Transmitter Telegraph Key Modulation Check

Test Equipment:
Telegraph key, oscilloscope, and dummy load if needed

Test Setup:
Couple the scope to the AM transmitter output circuits. Place a dummy antenna on the system, if the transmitter is not connected to its normal antenna.

Procedure:
Step 1. After you've completed the test setup, turn on the scope and transmitter (don't key the transmitter), and let them warm up.

Step 2. Next, key the transmitter and make any necessary scope adjustments that are required to display the signal on the face of the CRT. The width of the patterns may be set to any convenient length. The height will depend on the rf power output of the transmitter and coupling. Don't worry too much at this point about the scope sweep frequency.

Step 3. Key the transmitter with a series of dots. The only way you'll get successful results (evenly spaced and stationary scope patterns) is by using an automatic key. While observing the scope, set the scope sweep frequency to a slow rate and adjust it until you see a stationary pattern similar to the one shown in Figure 6-15.

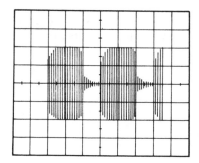

Figure 6-15: Oscilloscope pattern you should see from a plate-modulated transmitter, using a keying filter and keyed with an electronic key.

Step 4. Observe the scope patterns. They should be evenly spaced and not perfectly square (assuming there is a keying filter in the system). Ideally, each dot pattern should be rounded off at both the leading and trailing edges, as shown in Figure 6-15, not too much and not too little. Too much rounding off will cause the dots to run together and make it very difficult for the person at the receiving end to copy your message.

SSB Power Output Measurement

Test Equipment:
Oscilloscope, dummy antenna, and audio-signal generator (two audio signals may be used), see **Comments**

Test Setup:
See **Procedure.**

Comments:
The best way to check a single-sideband (SSB), class B amplifier for linearity and peak envelope power (PEP) output is with a scope. It is also a considerable help to have two audio tones available. But probably the most popular way to check the system is with a single audio oscillator plus some carrier. To add the carrier needed for the second audio tone, all that is required is to unbalance the carrier balance control in the SSB rig. The **PEP** measurement described here does not meet FCC requirements. You should consult your local FCC Field Office for information pertaining to the Commission's rules and regulations. However, FCC regulations require that the transmitter power be rated in terms of the dc input to the final stage.

Procedure:
Step 1. Place a dummy antenna on the SSB rig and couple the vertical deflection plates of the scope's CRT to the transceiver output. Or, as an alternative, use an rf probe and detector and couple directly to the scope's vertical amplifier input. To make the coupling, you can use either magnetic

coupling by using an rf pickup circuit (a couple of loops of wire placed near the transmitter output and connected directly to the scope's CRT vertical deflection plates), or a modified coaxial connector coupling tee such as the one shown in Figure 6-16. The tee connector is installed in-line between the SSB transceiver and dummy antenna. The scope CRT vertical plates are capacitive-coupled to the rf line, when connected to the straight adapter.

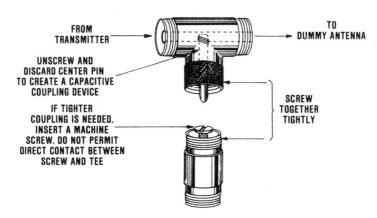

Figure 6-16: A capacitive coupler made using common coaxial connector components.

Step 2. It may be necessary to adjust the size of the scope pattern by moving the rf pick-up coil, or adjusting the machine screw in the capacitive coupler shown in Figure 6-16. Set the scope internal sweep for about 30 Hz.

Step 3. Set the SSB rig for normal voice operation and apply two sine-wave audio signals about 1,000 Hz apart (for example, 600 kHz and 1.6 kHz, or 3 kHz and 4 kHz), to the transmitter's audio input circuit. To do this, you can parallel two audio-signal generators, or use one audio-signal generator and some carrier, as mentioned in **Comments.** If you use this method, adjust the carrier unbalance control and signal generator output level control until you have equal amplitudes.

Step 4. Now, adjust the audio gain on the SSB rig and observe the pattern on the scope. You should see one of the patterns shown in Figure 6-17, depending on how you have the instrument controls adjusted and the equipment set up. The condition shown in (C) should disable the A/C system found in most rigs.

Step 5. Read the peak voltage (using an rf voltmeter is the easiest and most accurate way to do this). Next, convert your peak reading to rms by multiplying by 0.707. Then simply use the formula $P = E^2/R$. Simplified, all we have is this: $PEP = (E_{peak} \times 0.707)^2/R$, where R is the impedance of your dummy antenna.

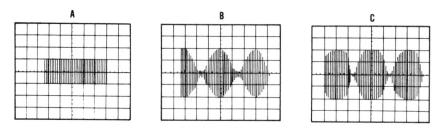

Figure 6-17: Pattern obtained by modulating an SSB transceiver by a single audio tone is shown at (A). The pattern at (B) is obtained by modulating an SSB rig with a combination of two audio tones. Serious distortion and splatter will result with a transmitter adjusted to produce the pattern at (C).

A Simple SSB Modulation Check

Test Equipment:
    Oscilloscope

Test Setup:
    See **Procedure.**

Comments:
    A measurement of percent of modulation for an SSB transmitter has no real meaning in reference to a measurement of an AM transmitter percent of modulation. In fact, the meter readings cannot be relied on for assurance that the transmitter is being properly modulated. The only way to be sure that the SSB rig is being operated within linearity limits is to use a scope and modulate the transceiver with your voice. However, the method explained under the heading SSB Power Output measurement is a very practical way to check modulation patterns. The limit of modulation on the SSB transmitter is set by the point at which *your voice* causes flattening of a typical SSB voice-modulated signal. It is important to realize that the modulation pattern shown is only instantaneous and would continuously vary in amplitude and pattern during normal speech (see Figure 6-18).

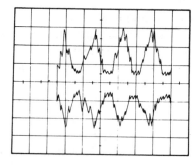

Figure 6-18: An example of what type of pattern may be seen on an oscilloscope during voice modulation of an SSB transmitter.

Procedure:

Step 1. Couple the vertical input of the scope to the transmitter rf output stage. The simplest coupling method is to use a piece of hookup wire, form a couple of turns at one end and connect the other two ends to the scope vertical input terminals. Place the pick-up coil near the transmitter's output stage and use the loosest coupling that will give a convenient height to the scope pattern.

Step 2. Set the scope to internal sweep. You'll need a slow sweep rate.

Step 3. Now, speak into the SSB right microphone and adjust the signal level for maximum possible height—one that doesn't show overloading. You want the best-looking pattern with no clipping, because clipping off the tops (flat topping) can cause serious distortion and considerable increase in spurious frequencies, causing splatter (see Figure 6-17 C.)

Step 4. After you have established the maximum possible height on the scope pattern without flattening of the peaks, mark this level on the face of the scope with a grease pencil, if you want to leave the scope hooked up during normal operation. Keeping the peaks just below your grease pencil marks will help assure you that you are not causing splatter and your signal will be free of distortion.

FM Percent-of-Modulation Measurement

Test Equipment:

AF signal generator, deviation meter, 50-ohm dummy antenna, and coaxial connecting cables.

Test Setup:

See Figure 6-19.

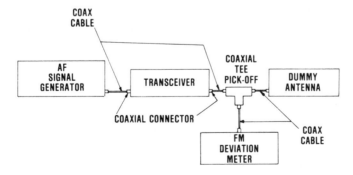

Figure 6-19: Test setup for making an FM modulation measurement.

Comments:

Refer to SSB Power Output Measurement for construction details of a coaxial tee pick-off attenuator.

Procedure:

Step 1. Assemble all equipment, as shown in Figure 6-19. Then load the transmitter into the dummy load.

Step 2. Set the AF signal generator to 1,000 Hz and adjust the FM deviation meter according to the manufacturer's instruction manual.

Step 3. Adjust your AF signal generator output level until you have a frequency swing of a predetermined value. The exact value of frequency swing for 100 percent modulation will depend on what is authorized by the FCC. For example, in FM broadcasting it is ±75 kHz and, if you're servicing certain mobile radios, it's ±5 kHz (ignoring permitted tolerance in both cases).

Step 4. To determine the percent of modulation, divide the reading by 100. As an illustration, let's say that you're checking a mobile radio where the maximum swing permitted is ±2.5 kHz on the deviation meter. Next, 2.5 kHz/5kHz × 100 = 50 percent modulation. To find the deviation ratio, simply divide the maximum authorized rf frequency swing by the maximum audio frequency that is authorized for the purpose of modulating the FM transmitter. Using another mobile radio example, 5,000/3,000 = 1.6 deviation ratio.

Step 5. Increase the AF signal generator output level while observing the deviation meter. If your reading exceeds the authorized swing (for example, ±5 kHz for mobile radio), the transmitter's deviation control must be readjusted until it prevents overmodulation.

Transmitter Frequency Measurements

Absorption Wavemeter Method

Test Equipment:
Homemade absorption wavemeter shown in Figure 6-20, mounted in a metal box.

Test Setup:
See **Procedure.**

Comments:
If you need to make a frequency measurement where accuracy is important—such as complying with the FCC—*this method is not recommended.* However, for rough checks of receiving equipment, transmitter tune-ups, and many other shop checks where accuracy is not of prime importance, this simple instrument and procedure will speed up troubleshooting and quickly isolate the fundamental frequency because it is very insensitive to harmonics (when loosely coupled to a tuned circuit). Two formulas that can be used in the design and construction of an air core coil are given below.

$$L = (r^2n^2)/(9r + 10D) \text{ and } n = \sqrt{L(9r + 10D)/r^2}$$

where L is in microhenrys, r is radius of the coil in inches, D is length of coil in inches, and n is the number of turns. Table 6-4 lists the construction details for four coils that can be used with the wavemeter shown in Figure 6-20, which will provide a frequency span of approximately 1 to 150 MHz. Other coils can be designed using the previously mentioned formulas.

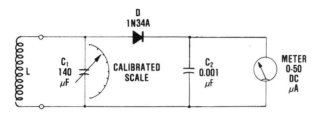

Figure 6-20: Schematic for a simple absorption frequency meter that, with suitable coils, will cover a broad range of frequencies.

APPROX. FREQ RANGE	VARIABLE CAPACITOR C = 140 pF
(A) 1-4 MHz	72 TURNS NO.32 INSULATED WIRE CLOSE WOUND ON 1" DIAMETER FORM
(B) 4-12 MHz	21 TURNS NO. 22 INSULATED WIRE CLOSE WOUND ON 1" DIAMETER FORM
(C) 12-40 MHz	6 TURNS NO.22 INSULATED WIRE ON 1" DIAMETER FORM. ADJUST TO LENGTH OF ⅜.
(D) 40-150 MHz	1 SHORT PIECE OF NO. 16 BARE COPPER WIRE. FORM A HAIRPIN LOOP WITH ½" BETWEEN STRAIGHT SIDES. MAKE TOTAL LENGTH, INCLUDING BEND, 2"

Table 6-4: Construction details for four coils needed to provide an approximate frequency span of 1 to 150 MHz, when used with the wavemeter shown in Figure 6-20.

You'll have to make a dial for the frequency meter and connect it to the variable capacitor. To prevent overcrowding of a single dial (if you want the full frequency coverage shown in Table 6-4), it's best to make a dial for each of the ranges shown. You can calibrate the instrument by using a signal generator or any other accurate source of rf energy. When you finish the circuit, place it in a metal box and take care during use that body capacitance does not affect your measurements.

Procedure:

Step 1. Select a coil that covers the frequency range of the equipment under test, connect it to the wave-meter-tuned circuit and place a calibrated dial on the capacitor shaft. Check your instrument calibration with a known-to-be-good signal source, if it hasn't been done recently.

Step 2. Energize the equipment under test and *loosely* couple the wavemeter coil to it. Tune the capacitor for resonance at the unknown frequency. Using very loose coupling insures that you are tuned to the fundamental frequency at a *maximum peak* (you'll probably find weaker peaks).

Step 3. After you are sure that you have the peak reading on the meter (you may have to vary the coupling and retune the capacitor a couple of times), simply read the unknown frequency off the wavemeter's homemade dial.

Frequency Counter Method

Test Equipment:

Frequency counter or frequency meter, coaxial tee pick-off (see SSB Power Output Measurement, for construction details), 50-ohm dummy antenna, and connecting coaxial cables

Test Setup:

Connect the equipment as shown in Figure 6-21.

Comment:

Transmitter frequency is best obtained by direct measurement. Usually, this is done by using a digital frequency counter. Typically, a low-cost counter will have an accuracy of 3 ppm at 25°C (this turns out to be less than 30 Hz at 10 MHz), while more expensive ones have 1 Hz resolution at 30 MHz. When

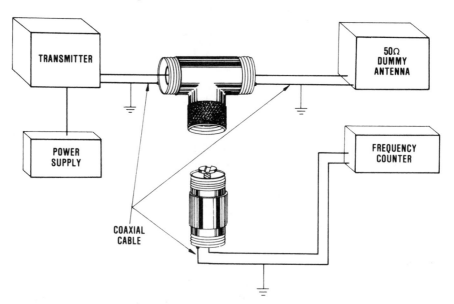

Figure 6-21: Test setup for making a transmitter frequency measurement using a frequency counter.

checking some transmitters, you'll have to use a counter that operates at much higher frequencies.

There are numerous frequency counters on the market, but the important point is that you must determine what frequency range and accuracy you need for servicing a particular transmitter. To do this, you must consult FCC rules and regulations (your local Federal Communications Commission Field Office can help you determine what government publications you need) for the service you are interested in (Public Safety Radio Service, Broadcast Service, etc.).

Procedure:

Step 1. Connect the equipment as shown in Figure 6-21 and wait until the radio is operating within specified temperature limits. Be sure the temperature is correct because any adjustments to the assigned frequency may result in an out-of-tolerance operation when the set is operated at the specific temperature.

Step 2. Key the transmitter and read the frequency off the frequency counter.

Step 3. Adjust to the assigned frequency, if necessary. Record the *actual* frequency in the log. Do this on all channels. *Note:* Although this adjustment can be done by anyone, it's *important* that the transmitter is checked by a person holding an appropriate FCC commercial License before it is connected to an antenna and used to radiate rf energy into the air (except for amateur radio operators, where an Amateur License is required).

SECTION 7

TV Servicing Tests and Measurements

SECTION 7.1: PRACTICAL GUIDE TO TODAY'S RECEIVERS

New ICs designed for use in solid state color television receivers never seem to end. For example, it's very possible you will encounter Motorola's MC1327 (dual doubly balanced chroma demodulator) in a late model color receiver. Figure 7-1 shows a typical application circuit using this IC, suggested by Motorola.

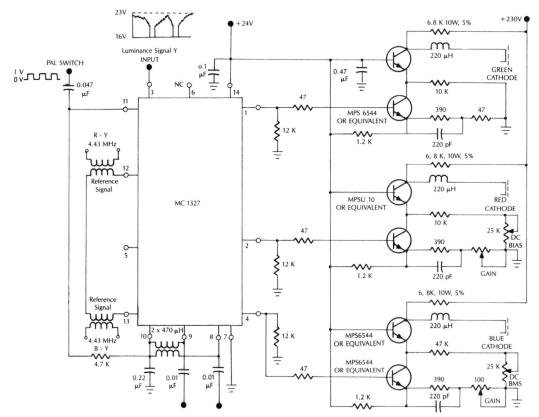

Figure 7-1: Typical applications circuit using Motorola's MC1327 IC.

You'll also find that black and white TV receiver ICs are in a state of change. In fact, the old black and white receiver circuits just don't exist in state-of-the-art receivers—a considerable amount of the system is contained in one IC. For example, Motorola's "Monomax black-and-white TV Subsystem" (MC13001), is a single chip IC that will perform the electronic functions of a monochrome TV receiver, with only the exception of the tuner, sound channel, and power output stages.

If you service a set using one of these ICs, it should be a decided improvement over the older receiver systems. For instance, the video IF detector is on-chip—no coils and no pins, except the IC input pins, of course. Also, the oscillator components are all on chip and no precision capacitor is required.

Although it is usually necessary to consult the TV set manufacturer's schematic and instructions when servicing a certain receiver, Figure 7-2 shows a typical application suggested by Motorola. The voltages listed should be in the ball park for just about any receiver using a MC13001P Monomax IC.

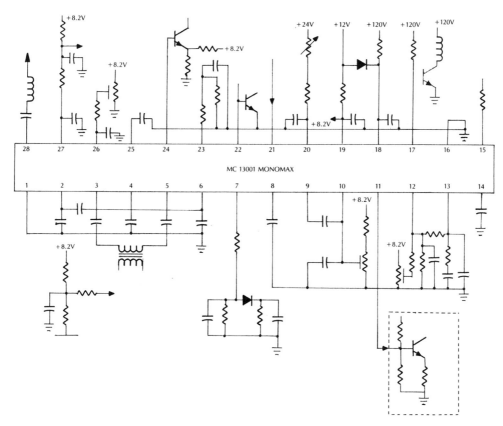

Figure 7-2: Typical application using Motorola's Monomax black and white TV subsystem MC13001P.

There are numerous other ICs that you will probably encounter when servicing modern TV sets. Some of these are: TV video IF amplifier (example, MC1352), TV sound amplifier (example MC13558), and complete sound system ICs such as Motorola's TDA1190P. This last device includes IF limiting, IF amplifier, low pass filter, FM detector, dc volume control, audio preamplifier, and audio power amplifier. Figure 7-3 shows the pin numbers and functions for this integrated circuit (TDA1190P or, TDA3190P).

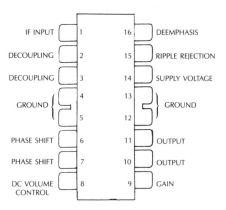

IF INPUT	1	16	DEEMPHASIS
DECOUPLING	2	15	RIPPLE REJECTION
DECOUPLING	3	14	SUPPLY VOLTAGE
GROUND	4 / 5	13 / 12	GROUND
PHASE SHIFT	6	11	OUTPUT
PHASE SHIFT	7	10	OUTPUT
DC VOLUME CONTROL	8	9	GAIN

Figure 7-3: TV sound system (TDA1190P or TDA3190P).

The TDA3190P is a 4.2-watt sound system designed for television and related applications. The TDA1190P is a low power version (1.3 watts). Incidentally, V_{cc} for the higher power IC is 24 V, load impedance 16 ohms. The low power IC uses a V_{cc} of 18 V and a load of 32 ohms.

SECTION 7.2: DISCRETE COMPONENT TV RECEIVER TESTS AND MEASUREMENT GUIDE

This section is fairly comprehensive, but stops short of requiring expensive professional-type equipment. Each test presents techniques that can be performed by a serviceman or technician with a minimum of low-cost test gear.

The VOM and its offsprings, such as the digital multimeter, are unquestionably the most popular and useful service instruments found in any shop. Therefore, you'll find that the majority of the tests included in this chapter use no more than a VOM and associated probes. Of course, some of the tests require the use of instruments such as a signal generator and VTVM, for instance. However, even these use the least expensive instruments available. On the other hand, some of the tests can be made without any instruments, and others need only a couple of electronic components that, more than likely, you already have in your spare parts box.

Picture Tube Video Signal Test

Test Equipment:
DC voltmeter and demodulator probe

Test Setup:
Remove the picture tube sockets and connect the voltmeter test probe to the signal input pin receptacle (the pin hole) in the socket. Then connect the voltmeter ground lead to a TV chassis ground point.

Comments:
The voltage reading you should expect will vary from set-to-set and also depend on the type of multimeter you use (VOM, VTVM, etc.). However, if the set is operating properly, your reading will probably be between 10 and 40 volts. For safety's sake, set your range switch to the 100-volt position to start, and then decrease it for an upper-scale reading, for best accuracy.

Procedure:
Step 1. With the TV set turned off, remove the back cover and disconnect the picture tube socket.

Step 2. Connect your voltmeter between the picture tube socket signal input pin hole and chassis ground. As a general rule, the color code for hookup wire is green for signal carrying wire and black for ground. *But not always,* so proceed with caution when using this premise.

Step 3. Energize the TV receiver and tune it to an operating channel. In many parts of the country, you can receive a test pattern on some channels. In most locations, you'll find it broadcast in the early morning or late at night (usually before 6 a.m. or after midnight). Generally, you'll get the most dependable results using a test pattern as a signal input during testing, particularly if you're comparing your readings to service notes. Of course, you can use a pattern generator if you happen to have one.

Step 4. If the receiver is operating normally, you should read a signal voltage of about 25 volts, ±5 or 6 volts, at the picture tube socket.

Ratio Detector Output Signal Test

Test Equipment:
High input impedance dc voltmeter with a demodulator probe

Test Setup:
See Figure 7-4.

Comments:
A below-normal reading may be caused by an improperly tuned system and the alignment should be checked if all other circuit components are

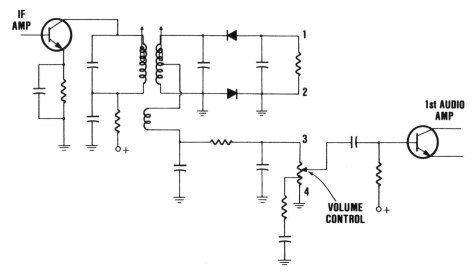

Figure 7-4: Typical solid state ratio detector circuit showing connection points for a demodulator probe

found to be good. *Do not try to do any aligning without proper equipment and instructions.*

Procedure:

Step 1. Connect the dc voltmeter demodulator probe across points 1 and 2 (these points will require a sensitive voltmeter), or points 3 and 4.

Step 2. Observe the voltmeter reading with the set tuned to a station transmitting a test pattern, if practical.

Step 3. If you're using a multimeter, such as a VTVM, with a full-wave demodulator probe, you should see a reading that compares favorably with the manufacturer's service notes. Other instruments, as a VOM, will probably produce readings that are somewhat lower than the manufacturer's specs. However, if you know what it should be under these conditions, a VOM can be used in place of a more expensive instrument.

Limiter-Discriminator Output Signal Test

Test Equipment:

High input impedance dc voltmeter with a demodulator probe

Test Setup:

See Figure 7-5.

Comments:

The discriminator shown in Figure 7-5 is merely an example of circuit design. In practice, you'll find variations between sets. A low voltage reading

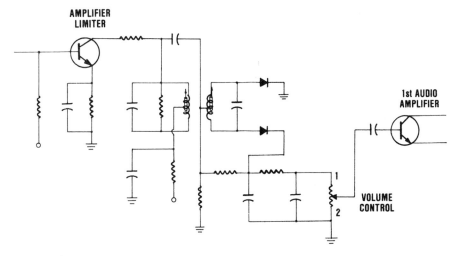

Figure 7-5: Example of a discrete component solid state limiter-discriminator showing connection points for a demodulator probe.

at points 1 and 2 in Figure 7-5 may result from improper alignment or trouble in the circuits preceding the limiter-discriminator. *Do not try to do any aligning without the proper equipment and instructions.*

Procedure:

Step 1. Connect the dc voltmeter demodulator probe across points 1 and 2.

Step 2. Observe the voltmeter reading with the set tuned to a station transmitting a test pattern, if practical (or use a pattern generator).

Step 3. Compare your voltage reading with the receiver service notes, or those obtainable from a receiver of the same type, in good operating condition.

Video Detector Test

Test Equipment:

DC voltmeter and high impedance probe

Test Setup:

Connect the voltmeter probe across the video detector load resistor. Figure 7-6 shows an example of a diode detector and associated circuits.

Comments:

Generally, you'll have little or no trouble with the video detector in a TV set of this type because there are no high voltages to contend with. However, if your voltage reading in the following procedure is zero, the first thing to do is check the diode.

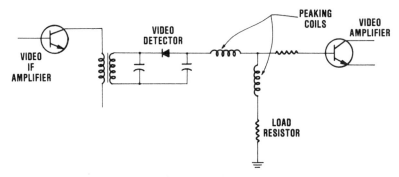

Figure 7-6: An example of a video detector system in a discrete component TV set.

Procedure:

Step 1. Connect the high impedance test probe of a VOM across the video detector load resistor.

Step 2. Tune the TV receiver to a station that is broadcasting a test signal, if practical (a pattern generator can be used).

Step 3. Observe the dc voltage reading. If you're tuned to a test pattern, the voltage reading should be steady. On the other hand, if you're tuned to a station broadcasting regular programming, you'll see the voltage reading fluctuating. In either case, a zero, or very low dc voltage reading, indicates trouble in the video detector or some preceding stage.

Video IF Stage Tests for Transistor and Tube TV Sets

How to Check an IF Stage Without Instruments

Test Equipment:
A 2-μfd, 50 working dc volts capacitor for checking solid state circuits, or a 0.02-μfd, 600 working dc volts capacitor for checking tube circuits, two alligator clips and a piece of hookup wire about a foot long.

Test Setup:
See **Procedure.**

Comments:
A dead IF stage can be found very quickly using the following procedure. Another use for the test setup is as a "field fix." For example, suppose an IF amplifier tube becomes defective during your favorite TV program. No problem. After pulling the tube, simply bridge the tube by placing a 0.02 μfd capacitor between the grid and plate connection. Naturally, the audio and picture will not be as good, but you may be able to see the end of the program, or get by for a few days. To make this test, or bypass an IF stage, you'll need a capacitor test lead, as shown in Figure 7-7.

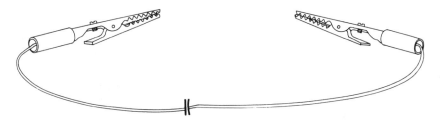

Figure 7-7: Capacitor test lead that can be used to check or substitute for an IF stage.

Procedure:

Step 1. With the TV receiver off, pull the last IF amplifier tube and connect one of the alligator clips of the capacitor test lead to the input of the IF stage and the other to the output.

Step 2. Turn the TV set on. If this stage is "dead," you should see a picture and hear sound—not as good as they were, but they should come in.

Step 3. If the sound and/or picture do not return, the last IF probably isn't defective. In this case, use the same method to check the other IF stages.

RF Signal Generator Method

Test Equipment:

RF signal generator modulated by a 400 Hz signal

Test Setup:

See **Procedure.**

Comments:

The signal generator can be any low-cost single-signal type capable of generating signals having frequencies up to the IF frequencies of the receiver under test.

Procedure:

Step 1. Set the signal generator to the IF frequency of the stage being tested. If, as is often the case, the IFs are stagger-tuned, the input circuit is not tuned to the same frequency as the output circuit. Figure 7-8 shows typical stagger-tuned IF frequencies.

Step 2. Connect the signal generator output leads to the input of the last IF stage. With the TV receiver operating, the 400 Hz modulation of the signal generator will produce sound bars on the screen if all stages after—and including—the last IF are working.

Step 3. Inject the signal into the input of each IF stage. When the signal generator fails to produce sound bars on the screen of the picture tube, you will have located the defective stage.

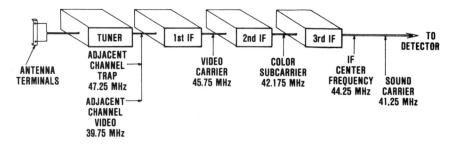

Figure 7-8: Typical stagger-tuned IF frequencies.

IF System Noise Level Test

Test Equipment:
   VOM

Test Setup:
   Connect the VOM leads across the TV receiver video detector load resistor. See Figure 7-6 for an example of a video detector system showing the load resistor. Also, short the TV receiver antenna terminals.

Comments:
   Higher-than-recommended dc voltage readings across the video detector load resistor indicate that the receiver will perform poorly in a weak TV signal area.

Procedure:
   Step 1. Short the TV receiver antenna terminals.

   Step 2. Connect your VOM across the set's video detector load resistor and observe the dc voltage reading, while the receiver is operating.

   Step 3. Compare your reading with the dc voltage reading taken from a TV set that is known to be in good operating condition. Or, consult the manufacturer's service notes for the recommended voltage reading. In any case, the lower the reading, the better.

VHF Receiver Front-End Test

Test Equipment:
   Any signal generator capable of tuning 55.25 MHz and being modulated with an audio signal either internally or externally. The signal generator's frequency accuracy is not of prime importance.

Test Setup:
   See **Procedure.**

Procedure:

Step 1. With the TV receiver turned on, set the receiver tuner to Channel 2.

Step 2. Set the signal generator to a frequency of about 55.25 MHz and apply an audio-modulated signal to the generator. It is very difficult—if not impossible—to accurately read the frequency setting on many low-cost signal generators. Therefore, you'll probably have to vary the frequency setting of the generator around 55 MHz to find the carrier 55.25 MHz. How to tell when you have it correct is explained in the following steps.

Step 3. Connect the signal generator test lead to the input of the TV receiver mixer. Now, tune your signal generator slightly below and above 55 MHz and watch for sound bars on the receiver viewing screen. The number of sound bars you will see will depend upon what audio frequency you're using to modulate the signal generator. You may have to vary the fine tuning control to get the sound bars. However, if sound bars appear, the mixer stage is operating.

Step 4. Connect the signal generator leads to the rf amplifier input and repeat the procedure given in Step 3. If you get sound bars again, this circuit is working.

Step 5. Change the signal generator frequency to the receiver's Channel 2 local oscillator frequency. Remember, in most cases, the local oscillator frequency is the sum of the incoming station frequency and the receiver's intermediate frequency. Also, don't use the audio modulation in this step or the next.

Step 6. Connect the signal generator in place of the local oscillator. If the TV receiver returns to normal operation (even for a second), it is an indication that the local oscillator is probably defective—assuming Steps 3 and 4 were found to be in working order.

AGC Voltage Tests

AGC Circuit Test

Test Equipment:
   VOM

Test Setup:
   Connect the VOM test leads between the AGC line and chassis ground.

Comments:
   The TV receiver service notes are not necessary (but they *are* helpful) to make general checks of AGC operation because, as you tune the TV receiver tuner, you should see voltage variations on the VOM, or the AGC system is not functioning.

Procedure:

Step 1. Set your VOM function switch to the dc position and connect it between the AGC line and chassis ground.

Step 2. Turn the TV receiver on and tune it to a no-station position on the tuner. Take note of the dc voltage reading.

Step 3. Next, tune the TV receiver to the strongest TV station in your area and, again, watch the voltmeter reading.

Step 4. Switch the TV receiver tuner back and forth between no-station and strong signal positions and you should see voltage variations occur. If you don't, the AGC system is not operating properly. Your next step is to check transistor and component parts.

AGC Problem at the RF Amplifier Test

Test Equipment:
   None

Test Setup:
   None

Comments:
   An AGC trouble in the RF amplifier will cause snow on the strong channels and the weaker channels may function properly.

Procedure:

Step 1. Disconnect the AGC line.

Step 2. Energize the TV receiver and tune it to a strong station.

Step 3. If the snow disappears, you have a problem in the AGC circuit.

Step 4. If the snow does not disappear (try all channels), check the first rf amplifier because, more than likely, it is causing the trouble.

Simple Vertical Linearity Test

Test Equipment:
   None

Test Setup:
   None

Procedure:

Step 1. Turn the receiver on and tune it to a station.

Step 2. Misadjust the vertical hold until you see the picture slowly start to roll (lose sync).

Step 3. As the picture rolls, you'll see a black blanking bar across the screen. By careful adjustment, you can get the blanking bar to hold fairly stable at the top and bottom of the screen.

Step 4. Check the thickness of the bar. It should have the same thickness at the bottom of the screen as at the top. If you see any variations in thickness, linearity of the set is off.

Checking for Sync Pulses

Test Equipment:

A 2-μfd, 50 working dc volts capacitor for checking solid state systems, or a 0.02-μfd, 600 working dc volts capacitor for checking tube-type sets, one alligator clip, a test probe, and about a foot of hookup wire.

Test Setup:

See **Procedure.**

Comments:

To make this test, you'll need to construct a capacitor test lead. All that is required is shown in Figure 7-9. *One note of warning:* When connecting to different points, be careful of strong signals. For example, output amplifiers may develop signals that are strong enough to damage the audio stages used in this procedure. In fact, it's better not to connect to any output amplifier when working on solid state chassis.

SEE TEST
EQUIPMENT FOR
CAPACITOR VALUE

Figure 7-9: Capacitor test lead that can be used to check for the presence of sync pulses.

Procedure:

Step 1. With the TV receiver off, connect the alligator clip of the capacitor test lead to the input of the audio output stage. Figure 7-10 shows an example of the connection in the sound section of a solid state color receiver.

Step 2. Turn the volume control to minimum.

Step 3. With the TV receiver operating, touch the test probe end of the capacitor test lead to the input, or some midpoint, of the sync section. An example of a connection point in a solid state TV would be the predriver or driver transistor.

Step 4. If the sync signal is present, you should hear a buzz in the speaker.

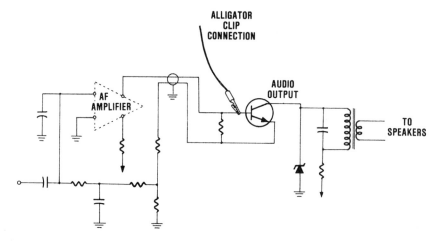

Figure 7-10. Test connections for using the sound section of a solid state color receiver for a signal tracer.

Simplified High Voltage Check

Test Equipment:
    Neon lamp voltage checker

Test Setup:
    None

Comments:

    If you simply want to see if there is a high voltage output, all you need is a number 51 neon lamp placed on the end of a wooden stick (dowel rod or old broom handle) about 10 inches long. See Figure 7-11.

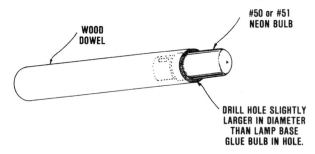

Figure 7-11: Construction details for a homemade high voltage checker.

Procedure:

    Step 1. To use this simple high voltage probe, place the neon bulb near a high voltage point (for example, the cap of a horizontal output tube).

Step 2. If there is a high voltage present, you'll see the bulb glow. Another thing you'll notice is that the closer the bulb is to the high voltage point, the brighter it will glow. With just a little practice, you can learn to make a pretty fair approximation of the value of any high voltage. For example, oscilloscope cathode ray tubes, closed circuit TV picture tubes, and automobile spark plugs, are just a few you can check with this tester.

TV Boost Voltage Measurement:

Test Equipment:
VTVM and a special high voltage dc probe that provides 100 times attenuation

Test Setup:
None

Comments:
Checking TV boost voltages can be a problem. Normal values of 1 kV often have a superimposed peak pulse of 1200 volts. A commonly occurring pulse form in the flyback circuit has a 15-kHz repetition rate and about 10 μ sec duration; sufficiently long enough to destroy your meter. The boosted B voltage in TV receivers is the voltage resulting from a combination of the B plus voltage from the power supply and the average value of voltage pulses (the 15 kHz, 10μsec pulses) coming through the damper tube from the horizontal deflection coil circuit. Although these pulses are smoothed by filtering, they are normally several hundred volts higher than the B plus voltage. One reason for the high voltage dc probe is to protect your VOM against these ac pulses. Another reason is that the length of the probe provides some protection for the user. *When servicing these probes, avoid handling the internal probe resistor! It is very susceptible to damage by moisture.*

Procedure:
Step 1. Turn the TV set on and attach the probe to the voltmeter. There are some special voltmeters mounted on probes that are built especially for this test which do not require the use of a VTVM. However, any good voltmeter with a high voltage probe can produce an accurate reading.

Step 2. Apply the probe between the point of measurement and the chassis ground (such as the high voltage connection on the picture tube, the plate of the damper tube, or plate of the horizontal output tube).

Step 3. Turn the brightness all the way down and note the high voltage reading. Remember, you're using a 100-to-1 attenuation factor, therefore, you must multiply the scale reading by 100 to determine the value of the dc voltage at the point under test.

Step 4. Turn the brightness all the way up and note the high voltage reading (again multiplying by 100, if you want to know the actual dc voltage at the test point).

Step 5. Compare the two readings and if there is considerable difference, it means the system is not operating properly.

Low-Voltage Power Supply Test

Test Equipment:
  VOM

Test Setup:
  See **Procedure.**

Comments:

The low voltage supply in a color TV receiver is similar to the low voltage supply used in black and white receivers; it furnishes the B plus and filament voltages (even all-solid state has picture tube filaments). However, a color receiver requires more power than a monochrome receiver. Therefore, you'll find a greater number of stages in these supplies. In fact, it isn't unusual to find 5 or 8 different value dc voltages in the output circuits of a color set's low power supply.

When you're making tests in solid state TV receivers, it's important to remember that the operation of transistors and other semiconductor devices can be seriously affected by slight voltage changes. The power supply voltage must be fairly constant and ripple-free. Therefore, power supplies in solid state TV receivers generally include a voltage regulator circuit.

Procedure:

Step 1. Measure the voltage across the output filter capacitor. Some examples of what you should read on some older TV sets are: 410 Vdc on an RCA CTC 16 chassis (this is the output voltage on the bridge circuit), 210 Vdc across one capacitor in a GE chassis CA, and 128 Vdc across the final filter capacitor of one of Zenith's solid state sets. If the voltage you measure is not up to the manufacturer's specs, check the diodes and filter capacitor (or capacitors).

Step 2. If the diodes and filter capacitors check out all right, test the power transformer secondary for possible shorted turns. You can do this with your VOM set to read resistance. Be sure to check between each half of the secondary (if the secondary is center-tapped). If there is an appreciable difference in reading, you probably have a partial, or total, short. *Do not forget to pull the ac plug and bleed those capacitors BEFORE making resistance measurements!*

Step 3. If the tests just described do not localize the trouble, check all other components and circuits such as the voltage regulator circuit, switches (there may be several switches, in some chassis), thermistor, and circuit breaker. You should measure about 1 volt across a thermistor after a short interval of operation (or measure approximately 120 ohms, when it is cold).

Of course, you should measure zero voltage across circuit switches and circuit breakers when the set is operating. A typical voltage measurement on the base of a regulator transistor (or the regulator output), should be about 24 Vdc. However, as always, either check a set of the same make or refer to the set's schematic.

Low-Cost Picture Tube Tester

Test Equipment:
Picture tube brightener

Test Setup:
See **Procedure.**

Comments:
The useful life of a TV picture tube usually is determined by the weakening of the electron emission in the beam from the oxide-coated cathode. Therefore, a safe, quick, and inexpensive way to check a picture tube that appears to be at the end of its life is to connect a picture tube brightener in the set. This will increase the power applied to the heater and raise the temperature of the cathode in the emission-depleted tube.

Procedure:
Step 1. Turn the set on, then darken the room. Even if you can only get a *very, very, faint,* glimmer of raster, the picture tube life can usually be extended. *Even if you don't see any light, continue this test.*

Step 2. Turn the TV receiver off and connect a TV tube brightener between the picture tube and its base socket. A picture tube brightener for your set can be purchased at almost any electronics supply store such as Radio Shack and, in most cases, comes with instructions.

Step 3. In many cases, the picture will snap in (sometimes, from absolutely nothing on the screen). If it does, the life of the picture tube may be extended by the simple addition of a brightener. TV picture tube brighteners for several of the most popular brands of TV sets in your area are handy things to have in your shop if you plan on servicing TV sets. Another trick is to place two brighteners in series to boost the filaments up another few volts (3 or 4 additional volts) for a short period of time, to help rejuvenate the tube.

Simple Picture Tube Filament Tester

Test Equipment:
Pilot lamp number 44 or 46, blue bead

Test Setup:
Remove the picture tube socket and insert the test lamp prods into the filament connections. See **Comments.**

Comments:

Sometimes, it is difficult to determine if the picture tube filaments are open, or whether it's just a bad socket connection. A quick, inexpensive filament checker can be made by soldering two insulated wires to a pilot lamp, as shown in Figure 7-12.

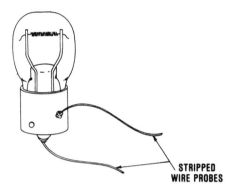

STRIPPED
WIRE PROBES

Figure 7-12: Low-cost TV picture tube filament checker. Although this illustration shows the wire connected directly to the lamp, a suitable lamp socket may be more convenient.

Procedure:

Step 1. Remove the picture tube socket and insert the test lamp leads into the filament connections. Connect them either way. It makes no difference because the filament voltage is usually an ac voltage (about 6 or 7 volts).

Step 2. If the lamp lights, the picture tube probably has burned-out filaments. If not, the socket probably is the source of the trouble. This simple little tester can save you a lot of time and trouble.

SECTION 8

Tests and Measurements for Antenna Systems and Transmission Lines

SECTION 8.1: HOW THIS SECTION WILL HELP YOU

In this section you'll find we stress making practical antenna measurements and tests with inexpensive equipment. Like many of the sections in this book this one has selected pieces of electronic test instruments that are easily home-built, and construction of this gear is described, or information is included that tells you where to find a schematic and parts list. As an example, you'll see how to make a quick systems test of a CB antenna—using a handy, homemade portable test antenna *that you can trust*. This test will get your servicing job off in the right direction—toward fast, accurate diagnosis and easy repair.

You'll find that the step-by-step instructions used in every measurement will help you speed through such testing procedures as determining antenna impedance, transmission line velocity factor, antenna front-to-back ratio, and more. Each one clearly maps out instrument setup and connections for easier measurements. Just go to the section that describes the exact test or measurement procedure for your particular need. Use the Table of Contents or Index to quickly find the one you're after. Each test is totally complete—or makes reference to another specific test, by test name. You'll find that the working illustrations and comments include all the information you need to make the test setup (or building instructions, if it's a homemade instrument), plus any additional shop tips that could help you skillfully conduct a successful and accurate diagnosis.

SECTION 8.2: ANTENNA TESTS

Mobile AM-FM Receiver Antenna Circuit Test

Test Equipment:
  None

Test Setup:
  Extend the antenna to maximum receiving height.

Procedure:

Step 1. Turn on the receiver and tune it to a very weak station, or noise, in the frequency range of 800 kHz to 1400 kHz (AM broadcast band).

Step 2. Next, adjust the antenna trimmer capacitor for maximum signal output. You'll frequently find that there is a small hole on the receiver case, near the antenna jack, that gives access to the antenna capacitor for the purpose of peaking. Figure 8-1 shows a typical electrical diagram of the antenna system.

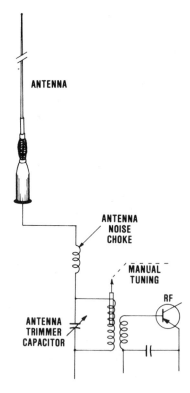

Figure 8-1: Mobile antenna circuit diagram showing antenna trimmer capacitor connections.

Step 3. If you can't find a definite peak, it's an indication there is a defect in the antenna system. Your next step is to see "Antenna System Leakage Test," the next test.

Antenna System Leakage Test

Test Equipment:
 Ohmmeter and jumper lead

Test Setup:
 See Figure 8-2.

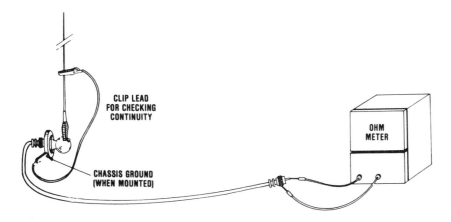

Figure 8-2: Test setup for testing an antenna system's leakage.

Comments:
 This test can be used to check out all types of antennas; mobile whip, vertical ground plane, TV, simple dipoles, etc. However, it cannot be used to check antennas using shunt feed or coupling networks that use impedance matching networks that present a dc short (for instance, the "delta" matching system).

Procedure:
 Step 1. Disconnet the antenna transmission line from the transmitter or receiver.

 Step 2. Connect a jumper lead across the antenna feed points, or between the whip and automobile chassis ground (rooftop, trunk lid, etc.) as shown in Figure 8-2.

 Step 3. Measure the leakage resistance of the antenna system with the ohmmeter. To do this, connect one ohmmeter lead to the center conductor of the coax (if the system uses a coax lead-in), and the other to the outer shield connector. In the case of unshielded 300-ohm TV lead-in, simply connect the leads across the two transmission line electrical conductors. The meter should indicate an open circuit. If leakage is detected, clean and check the insulator or, if you're checking a TV antenna system, check and/or replace the lead-in cable (assuming the antenna checks out okay).

Half-Wave Dipole Antenna Tuning Test

Test Equipment:
    Dip meter

Test Setup:
    See Figure 8-3

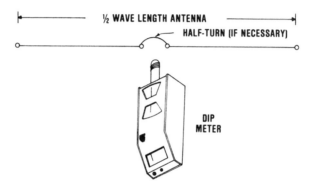

Figure 8-3: Illustration of how to couple a dip meter to a half-wave dipole antenna for the purpose of determining its loading frequency.

Procedure:

Step 1. Place the appropriate plug-in coil in the dip meter and set it to operate in the frequency range you expect the antenna to operate.

Step 2. Next, loosely couple the meter coil to the center of the dipole, using a short piece of hookup wire to create about one-half turn, if necessary (see Figure 8-3).

Step 3. Tune the dip meter for maximum dip and then read the frequency off the dip meter dial. *Note:* Generally, the accuracy of a dip meter is not exceptionally good.

Step 4. To improve the accuracy of your check, compare the dip meter to a calibrated communication receiver such as the ones used by SWLs, amateur radio operators, etc. *Note:* Be sure all guy wires and supporting structures are non-metallic (use plastic lines). Also, other antennas will have a pronounced effect on any antenna you're testing, if they are closer than one wavelength to the antenna under test.

Antenna Input Impedance Measurement

Test Equipment:
Dip meter, impedance bridge and microammeter

Test Setup:
See Figure 8-4.

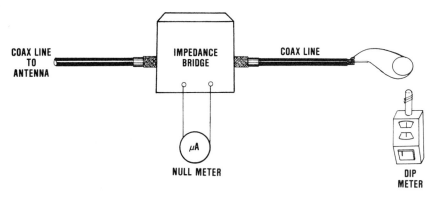

Figure 8-4: Test setup for measuring the input impedance of an antenna.

Comments:
There are several methods of measuring an impedance that is composed of both resistance and reactance. For instance, you can use a vector impedance meter such as produced by Hewlett-Packard Co., of Palo Alto, California, that has a readout in terms of polar coordinates ($Z = R\angle 0°$). That is to say, the readout is in terms of ohms and phase angle. Or, you can use the following procedure that has measured values that are of equivalent series form, $R \pm jX$.

Procedure:
Step 1. Although it can be done, in general, it's best not to use a transmitter to excite the antenna during this test because it's easy to burn up an RF bridge. Therefore, as a first step, disconnect the antenna from the transmitter.

Step 2. Connect the load input terminals of the RF bridge to the feed point of the antenna under test.

Step 3. Couple a dip meter to the impedance bridge, using the proper plug-in coil. Figure 8-5 shows a magnetic coupling coil and coax lead that you can make up, if needed.

Step 4. Adjust both the dip meter and bridge, trying for zero reading on the bridge null detector. If you can't get an absolute zero, it isn't all that important. However, it does mean the antenna isn't perfectly tuned. In other

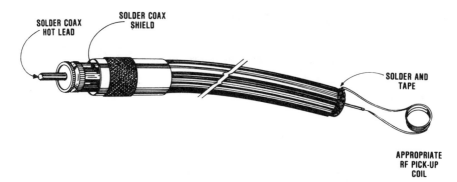

Figure 8-5: RF pick-up link assembly that can be used between an RF bridge and dip meter being used to excite an antenna for test purposes.

words, the antenna is acting as a slightly reactive load rather than a pure resistance. Trimming the antenna length or adding capacitance or inductance, as needed, should bring your reading closer to zero, if it's off considerably.

Step 5. After you have as near zero as possible, read the impedance of the antenna off the bridge indicating meter, and the antenna operating frequency off the dip meter.

Testing a Vertical Antenna with a Base Loading Coil

Test Equipment:
    Dip meter, impedance bridge, and special connecting cables (see **Procedure** for construction details)

Test Setup:
    See Figure 8-6.

Comments:
    Many commercially produced vertical antennas (especially for fixed station operation) have built-in resonant traps to make them broadband and also to make them appear electrically shorter. This test is not recommended for this type antenna because, in most designs, it does not have adjustable tuning coils. On the other hand, if you have a homemade vertical—such as a ground plane, etc.—using an adjustable base loading coil and you aren't sure it is adjusted for best reception or to produce the best radiation pattern, this test will give you a rough idea of whether antenna adjustments are needed.

Procedure:
    Step 1. Make up a coaxial cable one-half wavelength long (the cable on the left in illustration 8-6) at the operating frequency of the antenna. You'll

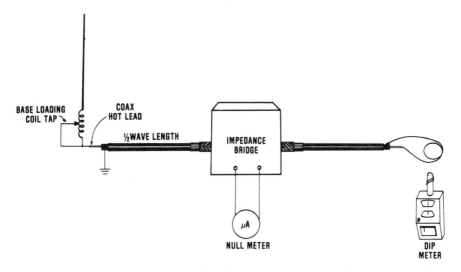

Figure 8-6: Test setup for checking vertical antenna with base-loading coil.

need only one coax connector at one end of the line. Connect the other end of the coax to the antenna, as shown, then make all other connections.

Step 2. Place the proper plug-in coil in the dip meter and adjust it to the approximate operating frequency of the antenna.

Step 3. Adjust the RF bridge, antenna base loading coil tap, and dip meter until you have the lowest reading possible on the null meter. Check the frequency by reading it off the dip meter. If it is too high, you need more inductance; too low, less inductance. Move your coil tap either up or down.

Step 4. The previous steps will let you know whether the antenna is properly tuned at the desired frequency, assuming the dip meter is accurate. To check the dip meter, you can zero-beat it against a calibrated communications receiver like those used by radio amateurs and SWLs.

Transmission Line Velocity Factor Measurement

Test Equipment:
Dip Meter

Test Setup:
See **Procedure.**

Procedure:

Step 1. First, you'll need to remove the insulation from a piece of coax and form a small loop (for coupling purposes) out of the inner conductor and then connect the inner conductor to the outside braid, as shown in Figure 8-7.

Step 2. Measure the exact physical length of the coax. Then use your calculator to determine the approximate *half-wave* resonant frequency of the line, using the formula:

$$\text{frequency (MHz)} = 468/\text{length (feet)}$$

Step 3. Next, set the dip meter to one-half your calculated frequency, couple it to the loop, and adjust the tuning dial for a dip on the indicating meter. Now, simply use your calculator with the following formula to find the velocity factor.

$$\text{velocity factor} = \frac{(\text{length in feet})(\text{dip meter frequency in MHz})}{246}$$

The formula is based on the formula for a quarter-wave stub, therefore, the figure for the dip meter frequency is one-half frequency of a half-wave line. The reason for this is that it is difficult to couple a dip meter to a coaxial cable with any accuracy. Using this method will help the situtation.

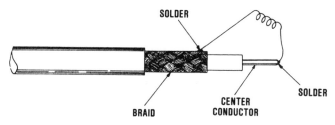

Figure 8-7: Homemade coupling coil for measuring transmission line velocity factor.

Antenna Front-to-Back Ratio Measurements

Field-Strength Meter Method

Test Equipment:
 Field-strength meter

Test Setup:
 See **Comments** and **Procedure.**

Comments:
 Whenever antennas are being checked, they should be installed at the proper distance above ground, and preferably in an open field. Ideally, antennas should be at least one wavelength above ground (for example, 36 ft. at CB frequency). When testing, install both antennas at the same height, and if guy wires are used, they should be non-metallic (plastic lines).

Procedure:

Step 1. Place the field-strength meter at least three wavelengths away from the directional antenna under test. Adjust the meter until its antenna has the same polarization as the antenna under test. *Note:* Complaints of poor front-to-back ratio can usually be traced to reflections from other antennas, buildings, or power lines, so be sure you are clear of such obstacles.

Step 2. Adjust the directional antenna under test so that it points exactly at the antenna of the field-strength meter. Energize the transmitter feeding the test antenna and then note the reading on the field-strength meter (E_1 in the formula in Step 4). Turn off the transmitter.

Step 3. Next, adjust the directional antenna to point the opposite direction and turn on the transmitter and take a field-strength reading (E_2 in the formula in Step 4).

Step 4. Now, because a field-strength reading is a voltage measurement, you can convert to dBs by using the familiar formula:

$$dB = 20 \log_{10} (E_1/E_2)$$

where E_1 is your first field-strength reading (Step 2) and E_2 is your second field-strength reading (Step 3).

Step 5. Once you've done Step 4, your answer is the front-to-back ratio of the antenna in dBs. One final word of caution. If you are using a homemade field-strength meter, be sure that the antenna lead-in of the meter does not pick up rf energy from the test antenna because this will cause your measurement to be erroneous. Also, it's best to use coaxial line to feed the test antenna, if possible, so that you won't have transmission line radiation. This will also cause your readings to be incorrect.

S-Meter Method

Test Equipment:
Receiver with an S-meter

Test Setup:
See Field-Strength Meter Method **Comments** and **Procedure** in this section.

Comments:
Depending upon the transceiver, an S-meter may be calibrated so that one S-unit is equal to 6 dB. Therefore, when using an S-meter of this type to perform a front-to-back ratio measurement, you'll find that an antenna has 1 S-unit gain between the front-to-back measurement and will also have a front-to-back ratio of 6 dB. *Caution:* Some S-meters are calibrated at only 3 dB per S-unit, and others at 3 or 4 at the low end and 6 or 7 at the top of the scale. *If in doubt, consult the manufacturer's service information.*

Procedure:

Step 1. Place a receiver (with S-meter) and its antenna at least three wavelengths away from the directional antenna under test. See Step 1 in the previous method.

Step 2. Same as Step 2 of the previous method except that you note the S-meter reading, using this method.

Step 3. This step is the same as Step 3 in the previous method except, again read the S-meter.

Step 4. Note the gain in S-units between Step 2 and 3. If 1 S-unit equals 6 dB, the gain in S-units multiplied by 6 will be the front-to-back ratio in dBs of the antenna under test. As explained under **Comments,** you can run into other methods of calibration, but the basic procedure is still the same.

SECTION 8.3: CHARACTERISTIC IMPEDANCE

Transmission Line Characteristic Impedance Measurement

Test Equipment:

VOM, impedance bridge, and dip meter

Test Setup:

See Figure 8-8.

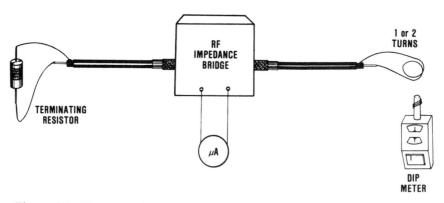

Figure 8-8: Test setup for measuring the characteristic impedance of a transmission line.

Procedure:

Step 1. Connect your equipment as shown in Figure 8-8. The resistor you use as a terminating resistor (load resistor) can be any value, to start. However, it should be a carbon type (i.e., non-reactive), having a value that you think is about the same as the characteristic impedance of the transmission line under test. The impedance of two-wire lines ranges from 100 to 300 ohms and the most often encountered coaxial line impedances are 50 and 75 ohms.

Step 2. Couple the dip meter to the test cable coil, as shown, and set it to the desired frequency. If you have selected a resistor of the same value as the line's characteristic impedance, you'll be able to adjust the bridge for a zero reading on the impedance bridge.

Step 3. If you can't adjust the null meter to zero, try another resistor and again attempt to bring the null meter to zero reading.

Step 4. When you have a zero reading on the bridge, try setting different frequencies on the dip meter. The null meter should remain at zero with changes of frequency.

Step 5. Measure the value of your final resistor; this is the characteristic impedance of the line under test.

SWR Measurement

Test Equipment:
In-line SWR meter

Test Setup:
See **Procedure.**

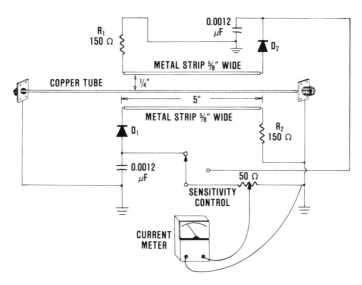

Figure 8-9: Typical in-line SWR tester design similar to the ones used by electronics manufacturers and home-builders. With components shown, the instrument works in 52-ohm transmission line at a power of about 80 watts or more.

Comments:

There are several popular low-cost directional coupler-type SWR meters sold in kit form and already assembled (for example, Heathkit and Radio Shack). However, it's very easy to make a home-built one. You'll find schematic diagrams in many publications written for amateur radio operators. Quite a few of them are designed around the schematic diagram shown in Figure 8-9.

The current meter shown in Figure 8-9 is usually a 0-100 microammeter. However, it can be a milliammeter if the transmitter output is strong enough. Just about any radio frequency diodes can be used for D_1 and D_2. For example, a matched pair of 1N34As will work well if the maximum forward current does not exceed 5 mA.

Procedure:

Step 1. Place the pick-up section of the SWR meter between the transmitter and antenna feed line. It is important that the impedance of the SWR meter matches the impedance of the transmission line. In most designs like the one shown in Figure 8-9, you'll find that the impedance of the instrument can be changed by using different values of resistance for the resistors labeled R_1 and R_2. For example, changing the resistor's value to 100 ohms instead of the 150 ohms shown will change the impedance of the unit to 75 ohms. See the manufacturer's instructions for other instruments of this type, to determine the value needed.

Step 2. Energize the transmitter and check that it is tuned to the correct operating frequency—peaking the tuning, if necessary.

Step 3. Check to be sure the SWR meter function switch is in the *forward position* and adjust the sensitivity control for a full-scale reading (if possible).

Step 4. Without making any other changes, place the function switch to the *reflect* position. If you're using a reflected power meter—such as Heathkit—simply read the SWR. But, if you're using a 0-100 μA meter in a homemade rig, you'll have to use the formula:

$$SWR = (10 + \text{reflected reading})/(10 - \text{reflected reading})$$

where 10 is full-scale reading. In other words, you're reading 100 μA as 10. For example, if you read 3 in the reflect position,

$$SWR = (10+3)/(10-3) = \text{about 1.85 to 1}$$

SECTION 8.4: TEST ANTENNA

Portable CB Test Antenna

Test Equipment:

Homemade coaxial antenna

Test Setup:
> See **Procedure.**

Comments:

The quality of a coax and connectors, and especially the soldering of coax to the connectors can affect SWR and gain. If you suspect a CB antenna system is using a poor grade of coax (this can cause a loss of 2 or 3 dB), or the system has poor soldering, a quick check can be run, using a portable antenna. Figure 8-10 shows the simple construction details for this antenna.

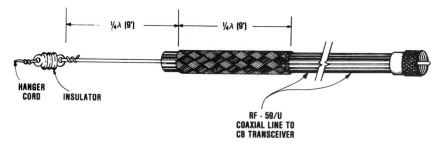

Figure 8-10: Construction details of a CB test antenna that can be taken almost anywhere. Cut off the outer cable insulation for a length of 9 feet and fold the shielded braid back over the cable as shown.

Procedure:

Step 1. Disconnect the CB antenna system from the transceiver.

Step 2. Connect your homemade antenna to the transceiver.

Step 3. String up the test antenna, attaching the insulator to some high point. Keep the test antenna as far as possible from other antennas, power lines, or any other material of a conductive nature. Ideally, nothing should be closer to your test antenna than 36 ft. (1 wavelength).

Step 4. Energize the transmitter and try to contact another CB station. If you are successful, you can be sure that the original antenna system needs replacement or repair.

SECTION 9

Practical Testing, Measuring, and Digital Instrument-Building Techniques

---

SECTION 9.1: TIMESAVING TESTS AND MEASUREMENTS USING A VOM

This section includes a collection of largely ignored testing techniques that can be done with ordinary low-cost gear. To illustrate, if your VOM doesn't have a low-voltage range, simply follow the procedure given under the heading "How to Use a VOM Low-Current Range to Measure Low Voltages." Want to check your VTVM for circuit loading? No problem. Turn to "Circuit Loading Test."

You'll also find step-by-step instructions for building, testing, and using three of the most important pieces of digital test gear; a logic memory probe, a logic pulser, and a logic monitor. These timesavers will be welcome additions to your workshop—and if you're thinking of starting a home service shop, they can give you the extra edge that is so important in the constantly changing electronics field.

Measuring Resistance in Megohms

Test Equipment:
DC voltmeter and dc power supply. Almost any dc source will do (25 to 600 volts), all you need is a measurable voltage.

Test Setup:
See Figure 9-1.

Comments:
Typically, the upper limit of direct resistance measurements with medium-priced multimeters will be 2, 100, or 1,000 megohms. However, you can extend the upper limit of resistance measurements by using the dc

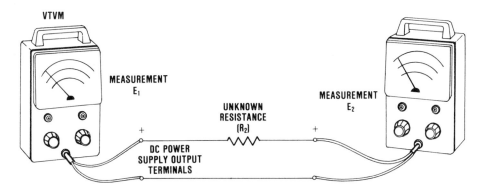

Figure 9-1: Voltmeter connections for determining the value of a resistor exceeding your ohmmeter's range.

voltmeter ranges of your multimeter, external dc supply voltage, and the simple formula

$$R_2 = R_1 [(E_1 - E_2)/E_2],$$

where R_2 is the unknown resistance value, R_1 is the dc voltmeter input resistance, and E_1 is the output voltage of the dc voltage supply, and E_2 is the voltage reading of the voltmeter when it is placed in series with the unknown resistance, as shown in Figure 9-1.

Generally, you'll find that high impedance voltmeters, such as a VTVM, will have an input resistance of 11 megohms with its dc probe attached (1 megohm in the probe). Therefore, the value of R_1 always will be 11,000,000 ohms in this case, or you can drop the zeros and use only the number 11, which will automatically give you an answer in megohms.

Procedure:

Step 1. Connect the circuit as shown in Figure 9-1 and measure the dc power supply voltage (E_1). Let's say you measure 385 volts.

Step 2. Measure the second voltage (E_2) by connecting the voltmeter's negative lead to the negative power supply terminal, and the voltmeter's positive test lead to the point shown in Figure 9-1. Let's say that you measure 1.5 volts this time.

Step 3. Compute the value of the unknown resistance, using the formula in the **Comments,** and assuming a voltmeter input resistance of 11 megohms,

$$R_2 = R_1 [(E_1 - E_2)/E_2] = 11[(385 - 1.5)/1.5]$$
$$= 2812.33 \text{ megohms.}$$

In round numbers, R_2 equals about 3,000 megohms.

VOM Moving Coil Meter Test

Test Equipment:

1.5-volt battery and resistor (see the **Procedure** for correct value)

Test Setup:
    See Figure 9-2.

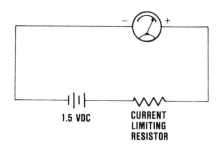

Figure 9-2: Circuit connections for checking a moving coil dc meter movement.

Comments:

If you suspect the meter in a VOM is defective, do not attempt to repair it yourself. After you have determined the meter is bad, using this test, return the VOM to the manufacturer. If there is evidence of acid core solder, paste fluxes, or irons cleaned with sal ammoniac, the VOM will not be served by most manufacturers. Instead, the instrument will be returned unrepaired. Also, when writing to the manufacturer, give a full description of the trouble, indicating which sections of the circuit you have tested. *Do not try* to make a continuity check of a meter movement with a standard ohmmeter because excessive current may damage the instrument.

Procedure:

Step 1. Calculate the value of current needed to provide about ⅔ full-scale deflection. For example, if the service manual shows the meter movement requires a current of 39.5 microamperes to move the pointer to full-scale deflection, ⅔ full-scale current is about 26.3 microamperes (⅔ × 39.5 μA = 26.3 μA).

Step 2. Calculate the value of series resistance needed to provide ⅔ full-scale deflection using 1.5 volts dc and a current of 26.3 μA. The value of the resistor, in our example, is about,

$$R = E/I = 1.5V/26.3\mu A = 57 \text{ k-ohms}$$

The nearest standard value is 56 k-ohms ±5 percent, which is close enough.

Step 3. Remove all connections on the back of the meter movement.

Step 4. Connect the resistor between the battery negative contact and negative terminal of the meter.

Step 5. Connect a jumper lead between the battery positive terminal and the meter positive terminal.

Step 6. When you make the last connection (Step 5), you should see a meter deflection very close to ⅔ full-deflection. If there is no deflection, or if the deflection is considerably above or below ⅔ full-scale, the meter is probably defective and should be sent in. However, it's best to go through your calculations one more time just to be sure that you are using the correct value series resistor in Step 2.

Circuit Loading Test

Test Equipment:
Resistor (preferably ±1%) having a value equal to the input resistance of the instrument being checked.

Test Setup:
First, connect the test instrument leads to the circuit under test and then insert the resistor in series with the test instrument test lead (see **Procedure**).

Comments:
This test is good when used with high input impedance instruments such as VTVMs, FETVMs and other electronic voltmeters where circuit loading should not be evident.

Procedure:
Step 1. Connect the voltmeter across the circuit to be measured. Record the voltage reading.

Step 2. Insert the resistor in series with the voltmeter (hot lead) and again measure the voltage at the same point you measured in Step 1. Record this second voltage reading.

Step 3. The voltage you measured in Step 2 should be almost exactly one-half the voltage you measured in Step 1. If the second voltage reading is greater than one-half, it's an indication of circuit loading. In other words, your direct measurements are in error.

How to Use a VOM Low-Current Range to Measure Low Voltages

Test Equipment:
VOM and a precision resistor (±1 percent)

Test Setup:
See Figure 9-3.

Comments:
Let's assume that your VOM has an internal resistance of 10,000 ohms for 50 μA (marked 0.05 milliampere, on some meter current range settings) full-scale deflection. Using Ohm's law, we find it takes 0.5 volts to produce this much current through a 10 k-ohm resistor. Now, if the test leads of the VOM *do not* have a resistor built in, we can use this range as a voltmeter having a 0.5-volt full-scale reading. In this case, simply read the 5-volt dc scale and divide by 10, or the 10-volt dc scale and divide by 20.

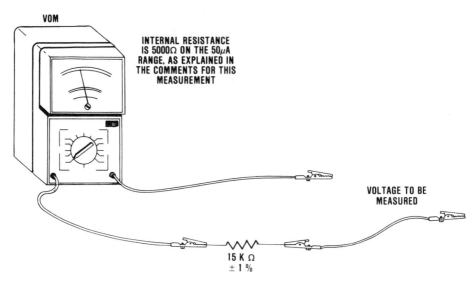

Figure 9-3: Test setup for converting VOM voltage and current ranges to measure low dc voltages.

If you want to provide a 2.5-volt full-scale range, you have to increase the total resistance of the 0.05 milliamp range to 50,000 ohms. Since the meter has a total internal resistance of 10,000 ohms on the 0.05 milliampere scale, we must insert 40 k-ohms in series with the VOM's test lead. Read the 2.5-volt full-scale on the 250 dc-volt scale and divide by 100.

Many VOMs have a voltage drop of 0.25 at full-scale, with an internal resistance of 5,000 ohms. Again, using Ohm's law, we find that 0.25 volts would have to be applied on the 50 μA range up to 20 k-ohms. Now, since the meter has an internal resistance of 5 k-ohms, you need to add 15 k-ohms in series with the VOM test lead. Read the voltage on the 100 Vdc scale and divide by 100.

Procedure:

Step 1. Set the VOM range switch to the 50 μA position, or any other current range you select.

Step 2. Consult the VOM operations manual to determine the internal resistance.

Step 3. Calculate the value of the series resistor needed, as explained in the **Comments** about this measurement.

Step 4. Insert the resistor in series with the VOM test lead and make the voltage measurement. Remember: *It's very easy to burn out a meter, so double-check your calculations before you do this step!*

Step 5. Read the voltage on the appropriate dc voltage scale and divide the reading by the correction factor (see **Comments**).

SECTION 9.2: OVERVOLTAGE PROTECTION CIRCUIT

All series pass transistors have to pass the total load current of the dc supply and, of course, can get very hot internally. This means that if you choose a marginal transistor for the job, it is very likely to short out prematurely and place the full output voltage of the bridge rectifier (or whatever type of rectifier you are using) on the main power bus of the equipment you are servicing. The scheme shown in Figure 9-4 can protect your equipment from this type of disaster. The circuit uses a high current SCR, and a fuse. In the event of excessive load current, the fuse is deliberately caused to open. This brute force approach is greatly speeded up by the *crowbar* action of the SCR.

Diode D_1 is a Motorola 5.6-volts zener (rated at 1 watt), and will not pass current until the output voltage reaches this level. Therefore, as long as the output voltage remains where you want it (at 5 Vdc), the fuse will not blow. But if the series pass transistor in your dc regulated supply shorts and the supply voltage shoots up (that is, up to 6.5 Vdc), then D_1 will conduct. This, in turn, permits current to pass to the gate of the 2N2573, causing the SCR to appear as a short circuit between the +5 Vdc line and ground and, of course, blows the fuse. This may seem like overprotection but it could save you a lot of components (and much money), especially when working with regulators tied to many dollars worth of ICs.

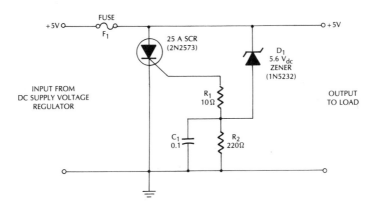

Figure 9-4: Overvoltage protection circuit (crowbar).

SECTION 9.3: USING A MC1741 OP AMP TO INTERFACE A HIGH IMPEDANCE TO A LOW IMPEDANCE

Comments:

The MC1741 general purpose operational amplifier requires no frequency compensation and is short-circuit protected. *Important:* An MC1741

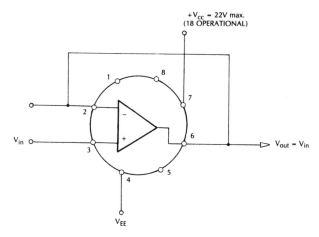

Figure 9-5: Wiring diagram for using a MC1741 OP AMP follower as an impedance-matching device.

IC, which is exactly the same type of OP AMP, has a *maximum* power supply voltage (V_{CC}) of only 18 volts on V_{CC} and V_{EE}. Typical *operating* voltage for the MC1741 shown in Figure 9-5 is 18 volts. Incidentally, even though this IC has provisions for offset voltage compensation (pins 1 and 5), we leave these pins unused in this example. The common mode input voltage range of this IC is typically ± 13 V.

SECTION 9.4: LAB-TYPE POWER SUPPLY FOR SHOP USE

Parts List:

C_1 Capacitor, electrolytic, 2200 μF, working volts 35 Vdc, axial. Radio Shack catalog No. 272-1020.

C_2 Capacitor, electrolytic, 100 μF, working volts 35 Vdc, axial. Radio Shack No. 272-1016.

D_1 Bridge rectifier, V_{RRM} volts 50. Motorola 3N246 or equivalent.

D_2 Zener diode, 9.1 volts. Motorola 1N4696 or equivalent.

Q_1 Transistor, NPN, Motorola 2N110 for a case power dissipation of 5 W @ 25°C with a maximum current of 1A. Our suggested maximum current load for this power supply is 250 mA.

R₁   Resistor, 560 ohms, ½ W.

T₁   Transformer. For about 9 Vdc output voltage, use a 12 V output transformer. Radio Shack catalog No. 273-1385 transformer may be used *if the load current does not exceed 300 mA.*

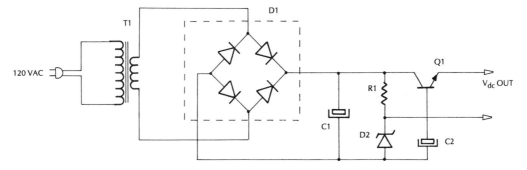

Figure 9-6: Wiring diagram for a full-wave bridge rectifier and active filter.

Comments:

The zener diode (D₂) and transformer you select set the output voltage of this power supply. For example, Radio Shack's transformer, catalog No. 273-1384, a 6.3 V transformer, and Motorola's zener diode 1N918A (5.1 V, 1.5 W) should produce the value of resistor R₁ when using these components.

SECTION 9.5: DIGITAL TEST INSTRUMENTS
THAT YOU CAN BUILD

Building a Logic Memory Probe

The logic memory probe schematic diagram shown in Figure 9-7 can be used to construct a simple but effective instrument that you can use to determine the logic condition of a digital circuit with pulse durations as short as 50 nano seconds.

You will need two ICs for this probe; the 7404 (a hex inverter) and an N8T22A (a retriggerable one-shot multivibrator). The Signetics N8T22A is a direct pin-for-pin replacement of the 9601 retriggerable one-shot. Therefore, either of these ICs will do the job. You will also need a transistor for the input. The one used (a general purpose 2N4401), is fairly inexpensive. The purpose of the transistor is to provide a high input impedance and to work as a buffer for the input to the hex inverter IC at pin 11. If you happen to have a spare parts box containing solid state components, you can use any transistor

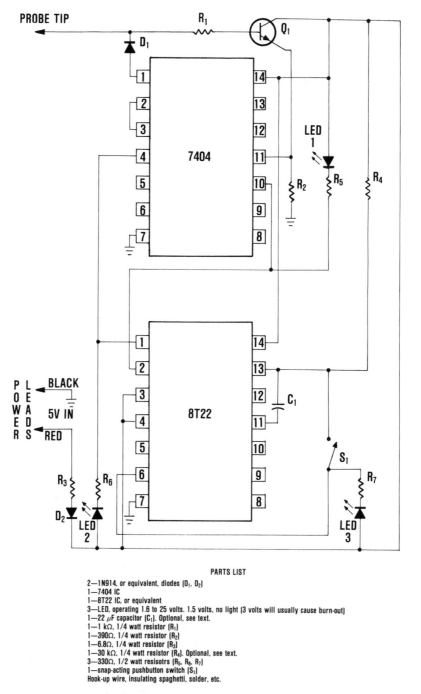

PARTS LIST

2—1N914, or equivalent, diodes (D_1, D_2)
1—7404 IC
1—8T22 IC, or equivalent
3—LED, operating 1.6 to 25 volts. 1.5 volts, no light (3 volts will usually cause burn-out)
1—22 μF capacitor (C_1). Optional, see text.
1—1 kΩ, 1/4 watt resistor (R_1)
1—390Ω, 1/4 watt resistor (R_2)
1—6.8Ω, 1/4 watt resistor (R_3)
1—30 kΩ, 1/4 watt resistor (R_4). Optional, see text.
3—330Ω, 1/2 watt resisotrs (R_5, R_6, R_7)
1—snap-acting pushbutton switch (S_1)
Hook-up wire, insulating spaghetti, solder, etc.

Figure 9-7: Digital memory probe circuit diagram. See parts list for component values.

with approximately the same electrical characteristics as the 2N4401. For example, a 2N222, usually found in VHF amplifiers and oscillators, will more than likely work.

Next, the three light-emitting diodes (LEDs). In most cases, it is necessary to limit the amount of current through an LED. Resistors R_5, R_6, and R_7 are current-limiting resistors. The 300-ohm value of these resistors will protect the circuit from excessive current flow and provide satisfactory operation of the LEDs.

Generally, the continuous forward current in LEDs is from 5 mA to 40 mA, and the forward voltage (V_F) drop of LEDs ranges from 1.65 to 2.2 volts. To calculate the approximate value of a current-limiting resistor, you can use the formula:

$$R_L = (V_{CC} - V_F)/I_F$$

where R_L is the value of current-limiting resistor, V_f is the voltage drop of the LED, and I_F is the forward current through the LED.

Comments:

In general, the value of a current-limiting resistor for an LED is not critical. After you have made your calculations, you'll probably have some oddball resistance value. You can vary from your calculated value quite a bit and still have satisfactory operation. For example, a couple of hundred ohms is acceptable although, in most cases, you are better off to use 330 ohms, which is more easily obtained.

For very compact construction, it's best to use a small capacitor for C_1. While they cost a little more, a dipped tantalum (electrolytic) capacitor is probably the smallest, in physical size, you can get. Whatever type capacitor you use, it should have a value of 22 µF, and a working voltage of at least 10 volts. A dipped tantalum capacitor made by Panasonic (stock No. TAC010), with a value of 22 µF, and rated for 25 volts, costs slightly over a dollar for a single unit.

You can use an external power source to operate the probe, or use alligator clips on the power leads and draw the operating power from the circuit you are testing. The purpose of diode D_2 and the resistor R_3 is to provide "wrong hookup" and overvoltage protection during use of the probe. The other diode, D_1, acts as a buffer for the transistor input. It protects the transistor from excessively high inputs and also helps maintain a high input impedance. The 1N914 diodes suggested, are fast-logic devices with the characteristics listed in Table 9-1. Any diodes with similar operating parameters can be substituted.

Capacitor C_1 and resistor R_4 control the time constant for the 8T22 IC, therefore, you can choose any two values that will provide a satisfactory operation within the limits of the circuit design. For example, 22 µF and 30 k ohms have a time constant of about one-half second and are what we have used in Figure 9-7.

DIODE	PEAK REVERSE VOLTS	MAX. FORWARD VOLTAGE AT MAX. mA	MAX. FORWARD mA AT MAX. V	MAX. REVERSE μA
1N914	75	1	75	0.025

Table 9-1: Operating characteristics of diodes (D_1, and D_2) shown in Figure 9-7.

An rf probe, penlight case, aluminum cigar holder, or any similar tube can be used for a case. The penlight case is probably the best because you can use the built-in switch. However, you will be very restricted in building room if you use a flashlight case. On the other hand, although the other type cases may give slightly more room inside for the circuitry, you will need to mount the switch (S_1) on the outside of the probe case.

The two alligator clips should be color-coded, one red and the other black, and have suitable length (about 18 or 20 inches of stranded cable). Zip cord (ordinary lamp power cable) will make a satisfactory power lead. Also, small-diameter coaxial cable can be used to obtain a more professional looking job.

How to Test the Probe

Step 1. Connect the alligator clips to a variable output dc power supply that is *set to 0 volts out.* Watch the polarity! Black to the negative terminal, red to the positive.

Step 2. Close switch S_1 by pressing the snap-action push-button switch. Now slowly adjust the dc voltage supply, starting from 0 volts and advancing toward 5 volts, all the while watching the LEDs. At 2.8 volts, you should see the memory LED flash on, then off, as you continue to increase the dc supply output voltage on up through 4.1 volts.

Step 3. Adjust the dc supply until its output is 5 volts.

Step 4. Touch the probe tip to the negative terminal of the power input cable (any ground point). You should see one LED light and another come on for just a short duration of time (about one-half second, with the components shown). The LED that stays on (during the time you are touching the common lead) is the low logic indicator. The LED that flashes on and off (memory LED) glows to indicate a positive/negative-going pulse. When you remove the tip from the negative voltage circuit, you should see the low logic indicator LED go out and the memory LED should glow again for about a half-second.

Step 5. Touch the probe tip to the positive terminal of the dc power supply. Again, you'll see the memory LED stay on for about one-half second and go off. However, you should see another LED come on and stay on. This LED is the high logic indicator.

Step 6. Your last step is to place switch S_1 into the other position (memory operation). Now touch the probe to either the negative or positive dc power source terminal. You should see the memory LED come on and stay on until you return the push-button switch back to the other mode of operation. In case you are a bit confused at this point, the memory mode (one condition brought on by pushing switch S_1) is designed to stay on, once it is triggered, until you cut it off. The stretch mode (the other position of switch S_1) is designed to observe pulses of short duration that the logic high LED and logic low LED can't catch.

How the Probe Works

If you do not have the probe tip in contact with any circuit (i.e., open), the input to the pin 11 of the 7404 IC is at a low logic and the other input, pin 1 of the 7404 IC, is at logic high (assuming you have the power leads connected to a 5-volt dc supply). LEDs 1 and 2 are off under these conditions, due to the action of hex inverter connections.

Now, let's say you touch the probe to a logic low (0.8 volts or less). This places pin 1 of the 7404 at a logic level 0 and, due to the wiring of the 7404, transistor Q_1 is cut off. During this time, LED 2 turns on, indicating a logic low level input at the probe tip. When LED 1 turns on (high logic), you'll find transistor Q_1 conducting and LED 2 off. Any sudden change of logic level at the probe tip will produce a negative-going pulse at the input of the 8T22 IC (pin 1 or 2), which, in turn, will trigger the multivibrator, resulting in LED 3 turning on for the time constant (set by C_1 and R_4) you have selected.

Constructing a Logic Pulser

Figure 9-8 shows a logic pulser schematic that is simple and easy on the pocketbook. The electronics can be housed in a hand-held probe. The pulser is programmed by pushing the switch (S_1) once for a change in logic level output.

A power cable with built-in probe protection should be constructed by using a short length (about 18 inches or so) of flexible cable and two protection diodes, as shown in Figure 9-9.

How to Test the Pulser

Step 1. Connect the alligator clips to a variable output dc power supply that is *set to 0 volt out.*

Step 2. Connect a dc voltmeter between the probe tip and the supply's negative terminal (ground).

Step 3. Slowly raise the variable dc power supply output voltage to +5 volts. Your voltmeter should read about +0.7 Vdc. The pulser is now in a logic low output condition.

Step 4. Press the push-button switch (S_1). Your voltmeter should jump to about +3.6 volts and the LED should come on.

If each of the four steps is performed and each voltmeter reading is fairly close to the ones given, your logic pulser is operating correctly.

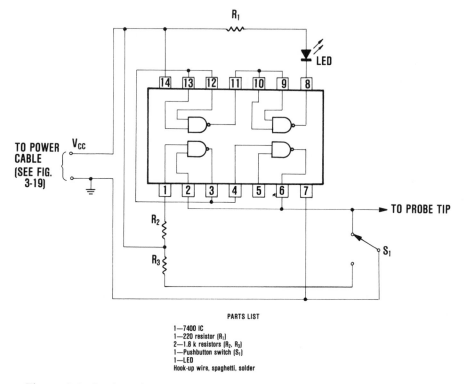

PARTS LIST

1—7400 IC
1—220 resistor (R₁)
2—1.8 k resistors (R₂, R₃)
1—Pushbutton switch (S₁)
1—LED
Hook-up wire, spaghetti, solder

Figure 9-8: Logic pulser schematic and parts list. V_{CC} = 5 Vdc for nominal operation.

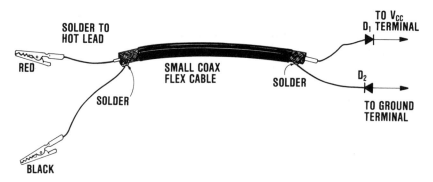

Figure 9-9: Power cable with instrument protection diodes for the logic pulser.

Incidentally, the procedure given is for testing positive logic. To test negative logic circuits (where a logic low input is required), simply keep the pulser switch pressed down and then release it for short pulses.

Building a Logic Monitor

A logic monitor (clip) has one advantage over a logic probe—it can be used to check for correct timing between a number of signals from the same IC. The logic monitor described allows *all* the pins of most ICs to be examined simultaneously, which means you will no longer feel helpless when faced with a timing problem.

Actually, the basic circuit for a logic monitor can be any one of the LED indicator circuits shown in Figure 9-10. However, C and D are particularly good because they will reduce the loading effect on the IC under test. Whether you use a bipolar transistor driver, Darlington transistor, or IC LED indicator circuit, it is important that you have as high input impedance as practical (1,000,000 ohms or more is best).

When you build your monitor, first you have to decide what number of pins you wish to check. For example, there are 14-, 16-, and more pin DIP IC packages. This is important because you must have 1 LED indicator circuit for *each* pin connection of the IC package you want to monitor.

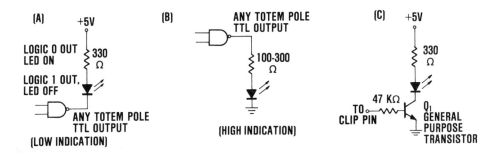

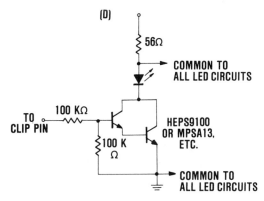

Figure 9-10: Basic LED indicator circuits. Operating power may be drawn from the IC under test.

14 DARLINGTON TRANSISTORS, MPS13, GE-64, TR-69, OR HEPS9100
14 LED's (RED)
14 RESISTORS (R₁)
14 RESISTORS (R₂)
1 RESISTOR (R₃) COMMON TO ALL CIRCUITS
1 14-PIN IC TEST CLIP (PROTO CLIP PC14)
1 14-WIRE RIBBON CABLE

Table 9-2: Parts list for a 14-pin logic monitor built using the Darlington transistor circuit shown in Figure 9-10.

Darlington Transistor 14-Pin IC Monitor

If you chose to use the Darlington transistor circuit shown in Figure 9-10 (D) and build a 14-pin DIP package IC monitor, you'll need the components listed in Table 9-2. If you want to build a 16-pin IC monitor, simply add two transistors, two LEDs, and two each of resistors R_1 and R_2 to the list. Add eight more of each of these components for a 24-pin monitor.

A 16-Pin IC Monitor Using ICs

The basic circuit and a parts list for a logic monitor using a 7406 hex inverter are shown in Figure 9-11. Figure 9-12 shows a 14-pin test clip and 14-wire ribbon cable with leads cut and bent for easy connection.

How to Test the Monitor

Step 1. Select a known-to-be-good IC (one in a properly operating circuit). Be sure you have the IC's recommended operating conditions, electrical characteristics, etc.; i.e., the spec sheets. The IC should have the same number of pins as your test clip.

Step 2. Connect the power leads from your logic monitor breadboard to a variable-output dc power supply, *set to 0 volts out*. You can use the circuit under test if you do not have a variable power supply.

Step 3. Place the test clip over the IC to be tested, making sure the clip is aligned properly. Align the clip mark (usually, a dot) with the index (dot, etc.) on the IC.

Step 4. Now slowly raise the variable dc power supply up to the IC's recommended operating voltage (usually about 5 Vdc). If there is any question about your wiring job, monitor the current out of the dc supply. A 7406 inverter is rated for 40 mA of current and the Darlington transistor is rated for 200 mA.

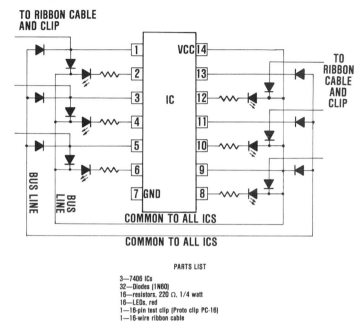

PARTS LIST

3—7406 ICs
32—Diodes (1N60)
16—resistors, 220 Ω, 1/4 watt
16—LEDs, red
1—16-pin test clip (Proto clip PC-16)
1—16-wire ribbon cable

Figure 9-11: Basic circuit and parts list that you can use to build a logic monitor around a 7406 hex inverter IC. Operating power is drawn from the IC under test.

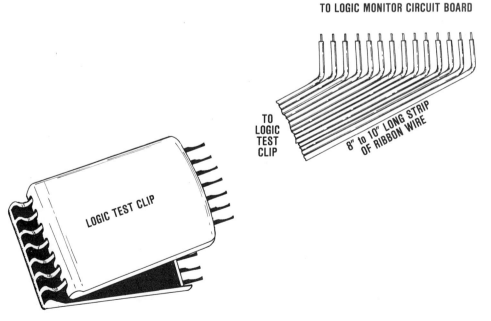

Figure 9-12: Example 14-pin test clip and 14-lead ribbon cable used when constructing a 14-pin digital monitor.

Step 5. Compare the ON and OFF LEDs with a truth table or other specs. Do this with several different ICs until you are sure that your monitor is reading properly. *Note:* Most 16-pin DIP packages use pin 16 as a dc source pin and pin 8 as ground, but not all. So, be sure you check which pin is V_{cc} and which is ground. You can only check ICs with the same connections for V_{cc} and ground as your clip. To check other ICs, you must change your monitor circuits.

When constructing the electronics, you can use a solderless breadboard, perfboard, or a "home-brew" breadboard of your own design. To see how the logic monitor using a 7406 IC works, let's assume pin 1 (Figure 9-11) is connected to the V_{cc} pin of the IC under test and pin 2 is connected to a ground pin on the IC. The diode at the base of the first LED on the left (LED 1) will conduct. Its bus line will go high. The other diode in the same circuit will not conduct. This keeps its bus line (off pin 1) isolated.

The first inverter inside the 7406 (pins 1 and 2) will see a high at its input, which will produce a low on its output, enabling LED 1 (the V_{cc} indicator) to turn on. The next inverter, pins 3 and 4, will see ground at its input (a low) which, in turn, will produce a high at the output and you will see the light (LED 2) off, indicating ground at the IC ground pin. All the rest of the IC pins will produce similar results, depending on whether the logic level is high or low. Since most 16-pin DIP packages use pin 16 as V_{cc} and pin 8 as ground, this means that, using this hookup, pin 16 will always glow and pin 8 will be dark.

If you decide to use Darlington transistors, as shown in Figure 9-10, you'll need 14, 16, etc., as we have said. The basic circuit works this way. The Darlington transistor drives the LED. Resistor R_2 is used to keep the input at logic level 0, to insure the LED will be cut off. Resistor R_3 is a current-limiting resistor needed by all LED circuits. However, if you use this circuit to build your logic monitor, you can use the resistor R_3 for a current-limiting resistor for all the Darlington transistor circuits you choose to use.

To save yourself a lot of trouble, be sure to mark each LED to show which pin and cable it belongs to. When breadboarding these circuits, it generally is less work if you'll first lay out one circuit (such as the Darlington transistor and its components shown in Figure 9-10) in one corner of the breadboard and work from there. Place your LEDs on one side of the perfboard and your wiring on the other side. To save space, mount your resistors vertically. Figure 9-13 shows how this can be done.

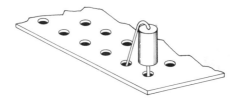

Figure 9-13: How to mount resistors vertically when breadboarding.

SECTION 10

Timesaving Tests and Measurements for Semiconductors

SECTION 10.1: HOW THIS SECTION WILL HELP SIMPLIFY SEMICONDUCTOR TESTS AND MEASUREMENTS

No professional "high precision" equipment is needed to perform any of the tests in this section. Each test has instructions for selecting the right low-cost test gear and applying it to the circuit (or component) properly to obtain the information needed. Furthermore, every test uses a step-by-step approach that will help you improve your ability to use inexpensive test instruments during your daily work.

The entire section is aimed at simplifying semiconductor tests and measurements. Solid state devices have a large number of parameters that can be tested—particularly, transistors. However, only a few of these are useful during actual servicing. The only thing most of us are interested in is determining if a semiconductor is defective, and then looking through our spare parts for a suitable replacement. Because this is all most of us really want to do, it's surprising how few pieces of test equipment are needed to check solid state components. Which procedure and test equipment you use will depend on what test gear you have on hand. To solve this problem, different ways of doing the same test are shown, with the simplest and least expensive way always given first.

SECTION 10.2: DIODE TESTING

Small Signal Diode Testing—Ohmmeter Method

Test Equipment:
  Ohmmeter

Test Setup:
  See **Procedure.**

Comments:

Generally, it's best to check the voltage levels at each range setting of your VOM, VTVM, FETVM, etc., i.e., the R × 1, R × 10, R × 100 ranges. Most VOMs will have 1.5 volts on the R × 1 range. In any case, when checking silicon diodes, you may need a supply voltage of 3 volts or more in order to cause the diode to conduct in the forward direction.

In most instances, when you switch your VOM to the ohmmeter function, the ground lead is positive and the red lead negative, but not in every case. Therefore, if you're not sure, it's best to check the lead polarities. Also, be careful with VTVMs because their polarity may be the exact opposite of VOMs.

Procedure:

Step 1. Connect the positive lead to the anode and the negative lead to the cathode. Your VOM should read a low resistance.

Step 2. Connect the negative lead to the anode and the positive to the cathode. You'll probably read infinity, or very near it, when checking silicon diodes. Germanium diodes should check out at several hundred k-ohms. A zero reading indicates the diode is shorted and should be discarded. The actual value of resistance measured means very little; the important thing is that a low resistance is measured in one direction and a high resistance in the other.

This test is a simple way to identify which end is the cathode and which is the anode, in the event they are not marked. Assuming the diode checks out to be good, Step 1 identifies the anode and cathode. However, small signal diodes come in various cases or packages, a few of which are shown in Figure 10-1. You'll notice that some types have the cathode end marked with a stripe or band. Others use a diode symbol on the metal case, as shown in Figure 10-1 C. Diode identification numbers may begin with 1N (for example, 1N60, 1N35, 1N4001, etc.), or start with letters such as GE.

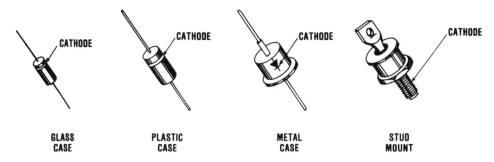

| GLASS CASE | PLASTIC CASE | METAL CASE | STUD MOUNT |

Figure 10-1: Diode packages.

Silicon Rectifier Peak Inverse Voltage Test

Test Equipment:

High voltage power supply, microammeter, potentiometer (about 500 k-ohms, in most cases), and voltmeter

Test Setup:

Connect the variable resistor, microammeter, and silicon rectifier in series, as shown in Figure 10-2. The silicon rectifier under test must be connected in reverse polarity for the PIV test. Connect the voltmeter in parallel with the diode.

Comments:

The most important parameters of a silicon power supply rectifier are the maximum-rated operating current and the peak-inverse-voltage (PIV). For example, you'll find that the specs for a General Electric GE-504-A silicon rectifier are 600 PIV and 1 amp forward operating current. The power supply used for the PIV test should be able to produce a voltage greater than the PIV rating of the diode under test. Incidentally, there are actually several PIV voltages: peak repetitive (in our example) and non-repetitive, which is usually higher (for example, 650 V). Also, there is a rms reverse voltage. In this case, it is usually lower, such as 450 volts.

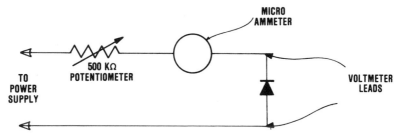

Figure 10-2: Circuit connections for a diode PIV test.

Procedure:

Step 1. Connect the circuit as shown in Figure 10-2 and set the potentiometer to full resistance.

Step 2. Turn on the power supply and adjust it to produce between 650 and 700 volts when testing GE-504-A. Many other silicon rectifiers are rated for a PIV of 400 volts. In this case, set your power supply to about 500 volts (see **Comments**).

Step 3. At this point, you may read a few microamperes, with only a slight increase as you slowly reduce the resistance. *Be Careful!* As you continue to reduce the resistance, you'll see a sudden sharp increase in current flow. When the meter begins to show this rapid rise, *quickly* note the voltage reading on the voltmeter and then shut off the power supply. The voltage you read on the meter just before the sudden current increase is the PIV rating of the diode. However, as we have explained, the actual circuit the rectifier is used in sets the highest PIV you can use.

Silicon Rectifier Forward Current Test

Test Equipment:

Low voltage, high current power supply, 200-ohm resistor with a high wattage rating (for example, 1 ampere through 200-ohms produces 200 watts of heat), ammeter, and voltmeter. In case the diode needs a heat sink and mounting, see Figure 10-3.

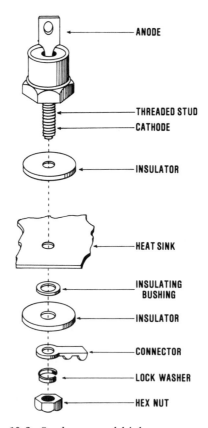

Figure 10-3: Stud-mounted high current rectifier.

Test Setup:

Connect the variable resistor, ammeter, and diode under test in series, as shown in Figure 10-4. Connect the voltmeter leads across the diode. Set the voltmeter to measure slightly over 1 volt. *Note:* The higher the diode current rating, the lower the voltage you'll measure across it; for example, about 1 volt for a 3-amp diode and 1.1 volt for a 1-amp diode. *Caution:* Lower current ratings may have higher voltages across the diode (up to 90 volts or more in some cases).

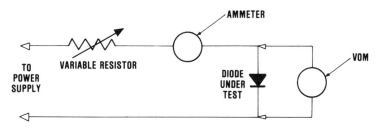

Figure 10-4: Test setup for checking the voltage drop, current, and power dissipated by a diode.

Comments:

When testing silicon diode rectifiers, it's fairly easy when checking small axial-lead packages such as the "top hat" type GE-504 used for an example in the preceding test. However, you can get into much higher current drains. Two examples of the high forward currents that you may encounter in these diodes are: 1) Hep-153, having a 200-volt PIV rating and a current maximum of 15.0 amperes, and 2) RCA SK03500, having a 600 volt PIV and a current maximum of 12.0 amperes. Typically, they are stud-mounted, as shown in Figure 10-3. The threaded stud usually is common to the case and electrically connected to the cathode. The anode is connected to the insulated terminal on the top.

Procedure:

Step 1. Slowly reduce the resistance value, watching both the ammeter and voltmeter.

Step 2. Generally, the current reading multiplied by the voltage reading shouldn't exceed slightly over one watt. However, in no case should this value exceed the wattage rating of the diode. If it does, the diode is assumed to be bad.

Step 3. Watch the voltmeter. It shouldn't read over 1 to 1.5 volts, in this example. Ordinarily, a silicon rectifier (such as this one) with a voltage reading over 1.5 volts should be discarded.

Silicon Controlled Rectifier Test

Test Equipment:

Ohmmeter and jumper lead

Test Setup:

See **Procedure.**

Comments:

When checking an SCR using this procedure, you many find some that won't respond to Step 3 because of insufficient ohmmeter current. However, some of them are very sensitive so, if in doubt, always use a high resistance range.

Procedure:

Step 1. Connect your ohmmeter between the anode and cathode with the gate open. See Figure 10-5.

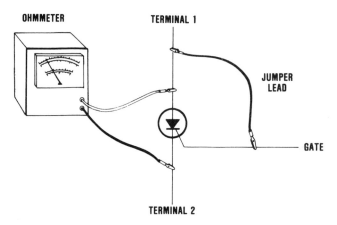

Figure 10-5: Ohmmeter connections for checking an SCR.

Step 2. Measure the resistance between anode and cathode. It should read very high. Reverse the ohmmeter leads and, if it's operating properly, you'll read the same value of resistance (practically infinite). Reconnect the ohmmeter leads, as explained in Step 1, before making the next step.

Step 3. With the ohmmeter still in the circuit, place the short (jumper lead) between the anode and gate. You should see the ohmmeter reading drop to a low resistance.

Step 4. Remove the jumper wire, without any other changes. The ohmmeter should show the same resistance reading as you had in Step 3—a low resistance. *Note:* Once the diode begins conduction, the gate loses all control.

Step 5. Disconnect and then reconnect either of the ohmmeter leads. The resistance reading should jump back to its original high resistance reading. This completes the check.

Practical Varicap Diode Test

Test Equipment:
Transistor tester or semiconductor analyzer

Test Setup:
Attach base and collector leads to the diode.

Comments:
Voltage variable capacitors, varicaps or varactors are PN junction diodes that perform like capacitors when biased in the reverse direction. Typically, the capacitance can be varied over a 10 to 1 range, with bias change from 0 to 100 volts and the current supplied by the bias supply not much more than a few microamperes, in most cases. One of the most important tests of a varicap is its reverse current leakage. This is an easy test to perform with a transistor tester.

Procedure:
Step 1. Connect the PN junction diode (varicap) to the transistor tester.

Step 2. Switch to both NPN and PNP positions. One will read forward and the other, reverse current.

Step 3. You should read no reverse leakage current on an ordinary transistor tester. Even the smallest reading indicates the varicap may be doing a poor tuning job and should be discarded.

Zener Diode Tests

In Circuit

Test Equipment:
Voltmeter

Test Setup:
None.

Comments:
The voltmeter reading across a properly operating zener diode will remain practically constant no matter how much current passes through the diode, provided it isn't driven beyond its operating range. There are two conditions that must be considered to successfully check a zener. First, there must be at least a few milliamps flowing through the diode to keep a stable voltage. Second, an excessive current flowing through it will burn it out. Just remember, the rated wattage of a zener is determined by the zener current at full conduction. Zeners come in a wide range of voltages—from a few volts to a few hundred volts. They are available in power ratings of 250 mW, 500 mW, 1 W, 10 W, 50 W, and a few other ratings not included. Therefore, when

making replacements, be sure to use an exact duplicate or, at least, use one that will stand up in the circuit you're working with.

Procedure:

Step 1. Measure the voltage across the load that is in parallel with the zener.

Step 2. If the voltage is quite a bit below the manufacturer's recommended value (or due to a comparison check, what you know to be normal), disconnect the zener. If your voltage reading jumps to well above the normal reading, it's an indication that the zener is leaking current and should be replaced. If there is no change, see Step 4. It's also possible that you'll read some voltage quite a bit above normal at all times. In this case, the zener is open.

Step 3. If you believe the zener is showing signs of an open circuit, disconnect it. If there is no change in the voltmeter reading, the zener is open.

Step 4. If your voltage reading is low when you measure across the load, it may mean an overload somewhere in the load circuit. Try disconnecting the zener and see if this changes the meter reading. If there is no change, you have an overload.

Out of Circuit

Test Equipment:

Power supply (this can be any bench power supply just so long as it provides a voltage that is a little higher than the zener-rated voltage), VOM, and two resistors, a 5- or 10-ohm potentiometer and a 1 k-ohm fixed resistor

Test Setup:

Attach the zener cathode, which may be marked plus, to the positive lead of the circuit, as shown in Figure 10-6. Next, connect the zener anode, which may be marked negative, to the negative lead. Attach the 1 k-ohm current limiting resistor in series with the zener and the potentiometer.

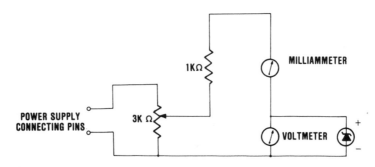

Figure 10-6: Test setup for checking the quality of a zener diode.

Comments:

Read last test **Comments** before making this test.

Procedure:

Step 1. Turn on the power supply and measure the voltage across its output. The correct voltage will depend on the zener under test. For example, the zener will not start operating until you reach about 20 percent of its maximum load. For instance, suppose it's a 20-volt supply and you're going to test a 12-volt zener. Your first check is to see that you have 20 volts out of the supply (a few volts one way or the other is close enough).

Step 2. Turn the power off. Set the VOM to read milliamperes. Attach the circuit as shown in Figure 10-6. Set the pot to minimum, turn power on and advance the pot, all the while watching the meter carefully. When the reading starts to rise rapidly, you have reached the zener's *breakdown voltage* (also called *zener knee, avalanche point,* and *zener voltage*). As you go up in voltage, the so-called *reverse current* will be extremely low until you reach the breakdown point, then the current will jump. If you increase voltage beyond this point, you should see only current increase. The voltage across the zener should remain constant. If it doesn't, the zener is bad.

How to Test a Tunnel Diode

Test Equipment:

VOM, variable high voltage dc power supply (2 or 3 hundred volts), and 20,000-ohm resistor

Test Setup:

See Figure 10-7.

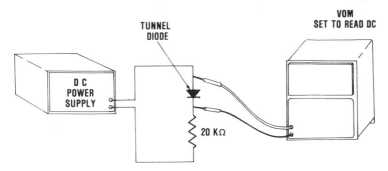

Figure 10-7: Test setup for checking a tunnel diode.

Procedure:

Step 1. Set the VOM to its lowest dc range. Typically, you'll be measuring about 0.5 volts.

Step 2. Adjust the dc power supply so it produces zero volts on its output and connect the circuit as shown in Figure 10-7.

Step 3. Adjust the power supply to produce an output voltage and watch the voltmeter reading. You'll see a sudden rise in voltage (from near zero to about 0.5 volts) when the diode switches, if it's performing correctly. *Note:* You may need several hundred volts out of the power supply to make the diode switch from a low to high voltage reading.

Step 4. Next, reduce the power supply output voltage until you again see the diode switch. You'll see the voltmeter that is connected across the diode drop back down to the minimum reading. If the diode switches to a higher voltage (about 0.5 volts) in Step 3 and drops to some minimum voltage value in Step 4, it's an indication that the tunnel diode is good.

SECTION 10.3: PRACTICAL TESTING PRECAUTIONS FOR SOLID STATE DEVICES

1. Take care when using a standard VOM because it can permanently damage the small electrolitics used in much of today's equipment.
2. Dropping solid state components such as FETs can damage them.
3. As a general rule, do not exceed maximum current and voltage levels *even temporarily.*
4. In most cases, it's best to install the solid state component last in the circuit. Complete all wiring and attach the dc supplies with switch off, before making any test.
5. When disassembling a test setup, switch off all voltages before removing the solid state component under test.
6. When soldering, use long-nose pliers to hold the component leads. Hold the lead until the solder is completely cool.
7. Many devices, such as the FET, are greatly affected by body capacitance that can cause errors in your testing. Always use a shielded probe and keep your fingers as far away as possible from the probe tip. *Don't* use an ordinary ohmmeter test lead held in your fingers, especially when working with operating FETs.
8. Ground your soldering iron tips and use a pencil type soldering iron.

SECTION 10.4: PRACTICAL TRANSISTOR TESTING

Ohmmeter Method

Test Equipment:
Ohmmeter. *Caution:* Although it seldom happens, small signal transistors may be damaged when tested in the following manner with an ohmmeter on the R × 1 scale.

Test Setup:

For high resistance checks, set the ohmmeter to the R × 1000 range. Use the R × 10 range when checking forward resistance values.

Comments:

This measurement provides a quick way of checking whether a transistor is good or bad. Also, the measurement will tell you whether a transistor is NPN or PNP. The test is based on the fact that a transistor responds to an ohmmeter exactly as two diodes back-to-back. Typical packages used for bipolar transistors are shown in Figure 10-8.

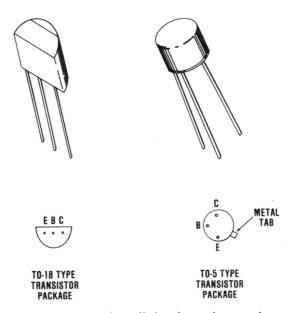

EBC

TO-18 TYPE
TRANSISTOR
PACKAGE

C

B METAL
TAB

E

TO-5 TYPE
TRANSISTOR
PACKAGE

Figure 10-8: Typical small signal transistor packages.

Procedure:

Step 1. Connect the ohmmeter between the base and emitter terminals (Figure 10-9 A and B). If the transistor is good, it should read low resistance in one direction and high in the other.

Step 2. Connect the ohmmeter leads between the collector and base terminals, and again it should read low resistance in one direction and high in the other.

Step 3. Using either polarity, connect the ohmmeter leads between the emitter and collector. You should measure a high resistance either way. A low resistance reading indicates the transistor has a leakage current and it probably should be replaced. These measurements will quickly identify a transistor as PNP or NPN (see Figure 10-9). It should be pointed out that

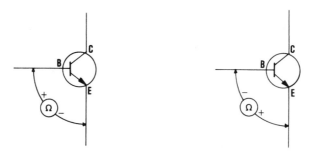

Figure 10-9: This diagram shows the polarities and resistances you should find when checking a good NPN transistor. A PNP-type transistor will be the exact opposite. It's important to know the actual polarity of the ohmmeter test leads in this test. If in doubt, check it with a voltmeter.

germanium power transistors have *normal* leakage current of up to 100 or 150 microamperes, or even more. However, silicon power transistors should have practically no leakage current; therefore, very high resistance readings when checking reverse directions.

Voltage Method—In Circuit

Test Equipment:
    Voltmeter (high input impedance), screwdriver or jumper lead, and 10 k-ohm resistor

Test Setup:
    See **Procedure.**

Comments:
    When making voltage measurements on transistor leads (common emitter or common base), you will find that a PNP transistor will have a positive voltage on its emitter, a negative voltage on its base, and a negative voltage on its collector. In other words, the base will be less positive than the emitter, and the collector should be the least positive of the three. However, in the basic NPN transistor, these polarities are reversed. The voltage between the base and emitter depends on the type of material used to manufacture the transistor. For example, you will find about 0.25 volts (anywhere from 0.1 to 0.4 volts) for germanium types (NPN or PNP), and about 0.6 (anywhere from 0.4 to 0.8 volts) for the silicon types. If your measurements are significantly different from these, the transistor is probably in need of replacement.

Procedure:
    Step 1. Connect the voltmeter across the collector circuit resistor (common emitter configuration) as shown in Figure 10-10.

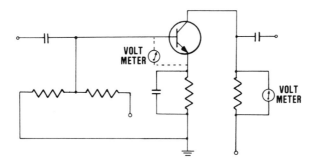

Figure 10-10: Voltage check points for a transistor amplifier.

Step 2. Short the emitter to the case with a screwdriver or jumper lead. If the transistor is operating properly, the voltmeter reading should increase. If the voltmeter shows no change, there's a problem in the collector-emitter circuit.

Step 3. Measure the forward bias. If it's low, or you measure no bias, connect a 10 k-ohm current limiting resistor between the collector and base (common emitter configuration).

Step 4. Monitor the collector voltage and place a 10 k-ohm resistor in the circuit. This should decrease the collector voltage reading if the transistor is good. If the voltage does drop, start looking for trouble in the bias circuit. See Figure 10-11.

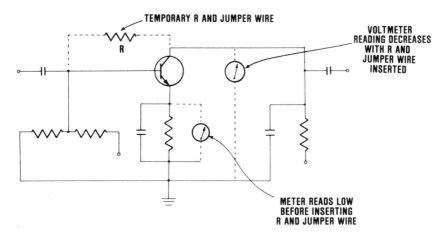

Figure 10-11: How to check a transistor by increasing bias.

Oscilloscope Method—In Circuit

Test Equipment:

Oscilloscope, 6.3 Vac filament transformer, 270-ohm resistor, 3 jumper leads and 2 test leads.

Test Setup:

Connect the transformer, resistor and test leads as shown in Figure 10-12.

Procedure:

Step 1. Remove all power supply voltages from the transistor circuit under test.

Step 2. Connect the tester to the equipment as shown in Figure 10-12.

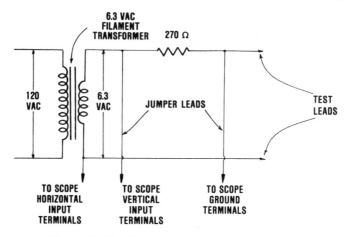

Figure 10-12: Transistor tester wiring diagram.

Step 3. Connect the test leads to each of the transistor leads. Reapply power. You should see a sharp right angle on the scope as shown in Figure 10-13. The waveform may be inverted or exactly as shown. Either case indicates that the transistor is good.

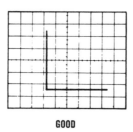

GOOD

Figure 10-13: Waveform for a good transistor junction.

Step 4. If the angle is rounded, the junction under test has leakage. A straight line means the junction is open or shorted. See Figure 10-14.

OPEN   SHORTED   LEAKY

Figure 10-14: Waveforms for a bad transistor junction. Reversing the test leads should cause the waveform to flip.

AF Power Transistor Collector Current Measurement

Test Equipment:
    DC milliammeter

Test Setup:
    See **Procedure.**

Comments:
    Many audio frequency power transistors are mounted as shown in Figure 10-15. Notice that spring clips are mounted to the underside of the heat sink and bolts screw into the clips. It's important that the mounting bolts do not short to the heat sink during normal operation or during the measurement to be explained. This type of mounting usually uses one of the mounting bolts to complete the collector circuit, as shown in Figure 10-15.

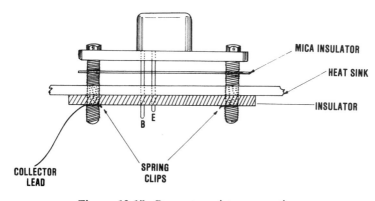

Figure 10-15: Power transistor mounting.

Procedure:

Step 1. Remove the bolt that has a terminal lug and wire between the clip and insulation. This is the lead that completes the collector circuit by permitting current to flow through the bolt to the transistor case, which is the collector in this type of transistor.

Step 2. Connect the milliammeter in series with the collector lead wire and the other transistor mounting bolt shown in Figure 10-16.

Step 3. Apply power and read the collector current. It is important to note that your reading is the collector current of this single transistor and does not include any other component currents that may be in the circuit during normal operation.

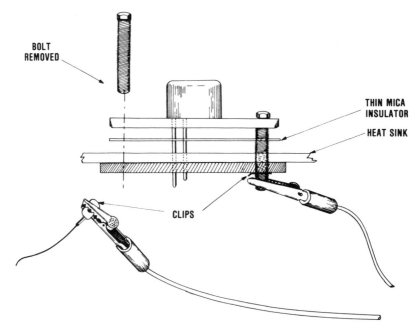

Figure 10-16: VOM Connections for measuring a power transistor's collector current.

Junction Field Effect Transistor Test

Test Equipment:
Ohmmeter

Test Setup:
See **Procedure.**

Comments:
Generally, when you purchase an FET, it comes with the leads twisted

together or it's wrapped in metallic foil. The leads should be shorted at all times the unit is not in use to prevent a static charge from building up on the gate. Incidentally, ordinary kitchen aluminum foil wrapped around an FET works very well for storage. When you're working with these devices, always be sure everything is grounded because even the static charge of your body may damage it—especially in some of the older MOSFETs.

Procedure:

Step 1. Connect the ohmmeter leads across the source and drain leads. You should measure a constant value from about 100 to 10 k-ohms. Reverse the ohmmeter leads and you should measure the same value of resistance. If you don't, the JFET is bad.

Step 2. Connect the ohmmeter leads between the gate and source. *Assuming an N-channel* JFET if the negative ohmmeter lead is connected to the source, you should read a very high resistance—almost an open circuit. You will find the exact opposite when checking P-channel JFETs.

Step 3. Connect the ohmmeter leads between drain and source. The gate-to-drain, or drain-to-source, resistance should be either low resistance or high resistance, depending on the ohmmeter lead polarity.

Dual FET Test

Test Equipment:
Ohmmeter

Test Setup:
See **Procedure** in JFET test.

Comments:
In dual type FETs, you'll find two source leads, two gate leads, and two drain leads. Test them in exactly the same way as in the preceding test, but check one at a time. If you find one section of the FET defective, probably the entire dual FET should be discarded although it's possible to use one section at a time, if you should want to save the half that is good.

Procedure:
See the **Procedure** given in the preceding test. Follow it for one-half of the dual FET and then repeat it for the other half.

Metal Oxide Semiconductor FET Test

Test Equipment:
Ohmmeter, two 1-megohm resistors (½ watt), and jumper lead

Test Setup:
See Figure 10-17.

Comments:
Unless the MOSFET is diode-protected, its leads should be short-

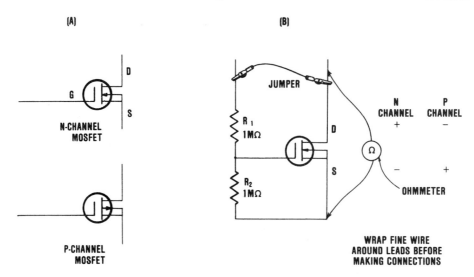

Figure 10-17: Test setup for checking an MOSFET (depletion mode shown).

circuited at all times except when installed in a circuit. Also, the transistor should never be removed from a circuit with power on since this may damage the unit. To disassemble a test setup, disconnect the ohmmeter and then remove the FET. It should be pointed out that although MOSFETs can be checked with an ohmmeter, it can be tricky because they are easily triggered by any stray transient voltage on the gate.

Procedure:

Step 1. Construct the circuit shown in Figure 10-17. Do all connections except the ohmmeter and shorting lead, then remove the shorting device, which may be a fine piece of wire wrapped around the FET pins.

Step 2. Connect the ohmmeter test leads across the source and drain. The negative lead is connected to the drain and the positive lead to the source, if the FET is a P-channel type. Use the reverse if it's an N-channel type—positive to drain, negative to source. (see Section 10-3, Practical Testing Precautions). The schematic symbols are shown in Figure 10-17 A.

Step 3. Note your ohmmeter reading. It can be anywhere from 100-ohms to 10 k-ohms for a depletion FET. An *enhancement* type should measure infinity.

Step 4. Short between the top of R_1 and drain with the jumper lead. You should see the resistance reading drop to a lower value when checking either type FET.

Step 5. Connect and disconnect the shorting lead several times and you should see the ohmmeter reading increase and decrease as you make and break the connection. If the rise and fall are quite large, the transistor is probably good.

SECTION 10.5: TESTING OPTOELECTRONIC DEVICES

Testing a Phototransistor Light-Operated Relay

Test Setup, (Parts List for Test):

C₁   Capacitor, 0.1 μF.

Q₁   Phototransistor, MRD300 or similar device.

Q₂   Bipolar transistor, MPS3394 or equivalent.

R₁   Resistor, 1.5 k.

S₁   Relay, miniature SPDT (high sensitivity) designed for transistor circuits with a low current drain. Contacts rated @ 1A @ 125 Vac, more or less, depending on the circuit the relay is to control.

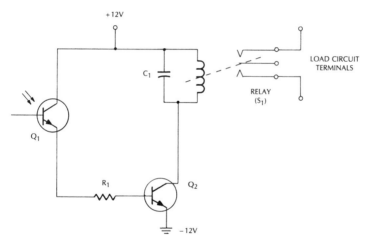

Figure 10-18: Wiring diagram for testing a light-operated relay using a phototransistor (MRD300).

Comments:

The base current of transistor Q_2 depends on the illumination of phototransistor Q_1. In turn, the value of the collector current (and, of course, the relay current) flowing through the transistor Q_2 is controlled by the Q_2 base current.

This current (Q_2 collector current) must be sufficiently large to activate the relay you are servicing. For example, if the relay requires 10 mA to close the points, Q_2's collector current should be at least 10 mA at saturation. *Note:*

The components chosen for this test must be rated to handle the relay current. The components used in Figure 10-18 should work with a relay rated at about 5 or 10 mA. But the biasing (current-limiting) resistor R_1 may have to be increased or decreased, if you encounter different solid state components and/or relay.

Procedure:

Step 1. After wiring the circuit as shown in Figure 10-18, apply the source voltage (about +10 or 12 Vdc).

Step 2. Energize the phototransistor. Illumination for the phototransistor should be fairly strong. The brighter the light source, the greater the current flow within the circuit.

Testing an Optoisolator (4N26)

Test Setup, (Parts List for Test):

IC_1 Opto coupler/isolator, 4N26 or equivalent.

R_1 Resistor, 47 ohms

R_2 Resistor, 2.2 k. *Note:* Collector current = 5mA.

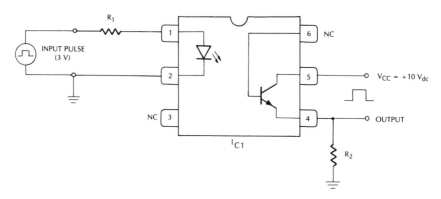

Figure 10-19: Wiring diagram for testing an optocoupler/isolator (4N26 or equivalent). See **Comments.**

Comments:

You can place an LED in the circuit between pin 2 and ground to monitor the input pulse. The output will pulse either in phase with this LED, or out of phase, depending on where you place the resistor R_2.

If you place R_2 in the transistor emitter circuit (as shown), the output will be in phase with your input signal. Or place R_2 in the transistor collector circuit and you should find that the output is 180° out of phase, in respect to the input.

Larger input signals may be used or required, depending on what type of optoisolator you are testing. However, resistor R_1 will have to be increased in value to insure that the diode is not subjected to currents over its rating. Refer to the coupler spec sheet for the current rating of the diode in the coupler you are servicing. Also, additional amplification of the input signal may have been achieved by adding a transistor amplifier and connecting its input (through a capacitor) to the output of the coupler (at pin 4).

Procedure:

Step 1. After wiring the device as shown in Figure 10-19, apply the source voltage to pin 5 and ground.

Step 2. Apply an input pulse and you should see an in-phase output pulse (using the wiring diagram shown in Figure 10-19. See **Comments.**

SECTION 10.6: HOW TO TEST TRIACS

Test Equipment:

Ohmmeter and jumper lead

Test Setup:

In most cases, start off with your ohmmeter set to the R × 1 range. However, it's possible that the R × 1 range will produce too much current for a sensitive triac. To eliminate the problem, place 10 to 15 ohms of resistance in series with the ohmmeter test lead. On the other hand, you may encounter triacs that won't trigger with an ohmmeter. You'll need a larger voltage than the ohmmeter can produce to check these. But they should respond in exactly the same way as described in the following procedure.

Comments:

Basically, the triac is a semiconductor-type switch with three electrodes—two main terminals and a gate. For all practical purposes, it is equivalent to two SCRs connected in parallel (with one inverted), with a common gate. It provides switching action for either polarity of applied voltage and can be controlled in each polarity from the single gate electrode.

Procedure:

Step 1. Connect the positive lead of your ohmmeter to one of the main terminals (T_1, T_2) and the negative lead to the other, as shown in Figure 10-20. The ohmmeter should read a very high resistance.

Step 2. Momentarily short between either terminal 1 or 2 and gate with a shorting lead. The ohmmeter reading should immediately drop to a much lower reading. Remove the shorting lead and there shouldn't be any noticeable change in resistance. This completes the test on one-half the triac.

Step 3. Connect the shorting lead between the gate and the other terminal. You should again read a very high resistance.

Step 4. Short the same two elements (gate to terminal) you used in Step 3 and the resistance should drop to a low reading. Remove the shorting lead and the ohmmeter reading should remain the same. This completes the test on the other half of the triac.

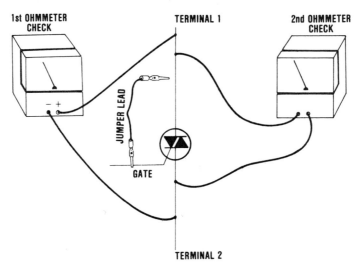

Figure 10-20: Triac test connections.

Index